EFFICIENCY, SUSTAINABILITY, AND JUSTICE TO FUTURE GENERATIONS

Law and Philosophy Library

VOLUME 98

For further volumes:
http://www.springer.com/series/6210

EFFICIENCY, SUSTAINABILITY, AND JUSTICE TO FUTURE GENERATIONS

Edited by

Klaus Mathis

University of Lucerne, Switzerland

Springer

Editor
Prof. Dr. iur. Klaus Mathis, MA in Economics
University of Lucerne
Faculty of Law
Frohburgstrasse 3
P.O. Box 4466
CH-6002 Lucerne
Switzerland
klaus.mathis@unilu.ch

ISSN 1572-4395
ISBN 978-94-007-3768-6 ISBN 978-94-007-1869-2 (eBook)
DOI 10.1007/978-94-007-1869-2
Springer Dordrecht Heidelberg London New York

Printed on acid-free paper

Springer is part of Springer Science+Business Media (www.springer.com)

Preface

This anthology, "Efficiency, Sustainability, and Justice to Future Generations" arises from the Special Workshop of the same name which took place at the 24th World Congress of Philosophy of Law and Social Philosophy in Beijing, 15–20 September 2009. The thematic scope of the volume spans both "Law and Economics" and "Law and Sustainability". Since the majority of the essays are by authors from Europe, they reflect a specifically continental-European perspective on these themes. One of the main intentions behind the publication of this volume, therefore, is to make this point of view accessible to an English-speaking readership.

I take this opportunity to thank all the people who have contributed to the successful completion of this book. First of all, I thank Deborah Shannon for her usual meticulous translation of all my essays, as well as those by Paolo Becchi and Balz Hammer. I also thank Lynn Watkins and Claudio Staub, BLaw, for their diligent proofreading of the entire volume. Their assistance would not have been possible without the generous support of the Dean's Office of the Faculty of Law at the University of Lucerne. Special thanks, therefore, go to the Dean, Prof. Dr. Regina E. Aebi-Müller, and the Faculty Manager, Attorney at Law lic. iur. Marcel Amrein.

Finally, I am grateful to the editors of the "Law and Philosophy Library" series for kindly including this volume in their series, to Neil Olivier, Diana Nijenhuijzen and Corina van der Giessen at Springer Publishers for overseeing the publishing process, and to Anandhi Bashyam and her team from Integra Software Services for the careful typesetting.

Lucerne, Switzerland
July 2011

Klaus Mathis

Introduction

Fifty years after the famous essay "The Problem of Social Cost" (1960) by the Nobel laureate Ronald Coase, Law and Economics seems to have become the *lingua franca* of American jurisprudence, and although its influence on European jurisprudence is only moderate by comparison, it has also gained popularity in Europe. A highly influential publication of a different nature was the Brundtland Report (1987), which extended the concept of sustainability from forestry to the whole of the economy and society. According to this report, development is sustainable when it "meets the needs of the present without compromising the ability of future generations to meet their own needs".

A key requirement of sustainable development is justice to future generations. It is still a matter of fact that the law as well as the theories of justice are generally restricted to the resolution of conflicts between contemporaries and between people living in the same country. This in turn raises a number of questions: what is the philosophical justification for intergenerational justice? What bearing does sustainability have on the efficiency principle? How do we put a policy of sustainability into practice, and what is the role of the law in doing so?

The present volume is devoted to these questions. In Part I, "Law and Economics", the role of economic analysis and efficiency in law is examined more closely. Part II, "Law and Sustainability", engages with the themes of sustainable development and justice to future generations. Finally, Part III, "Law, Economics and Sustainability", addresses the interrelationships between the different aspects.

In the first essay, "Consequentialism in Law", Klaus Mathis analyses the significance of economic analysis methods in law. It is shown that the admissibility of economic analysis methods in the application of law is closely linked to the question of whether and to what extent consequentialist arguments are admissible into juristic argumentation, i.e. whether any of the consequences of his decision are legally material for the judge, and if so, which ones.

One of the fiercest opponents of consequentialism in the application of law must be Niklas Luhmann. His principal arguments against it are the following: legal certainty, legal equality, the overburdening of the courts, and the threat to their independence. Despite a longstanding tradition of consequentialism in Anglo-American law, it is not without its critics. One of the most eminent is Ronald Dworkin who, in "Hard Cases" (1975), makes a distinction between "principles" and "policies", the latter being arguments in the interests of the common good or collective goals, and argues that courts, when deciding on hard cases, should essentially confine themselves to arguments of principle, because political arguments about objectives are the preserve of the legislator.

In the German-speaking world, a differentiated view seems predominant today. Thus, Ulfrid Neumann rightly talks about the "Janus headedness" of legal argumentation: on the one hand, it is highly authoritarian in outlook, but on the other hand, he notes a growing tendency towards consequentialistic argumentation, i.e. with reference to the likely impacts of the judgement. Mathis concludes that whenever those who apply the law become makers of law (*modo legislatoris*), they are subject to the same responsibility for consequences as the legislator, a responsibility that is only fulfilled if they carry out the relevant impact assessments. This applies specifically to judicial development of law and – in hard cases at least – even to the interpretation of legal norms.

Similarly, the second essay, "Consequence-based Arguments in Legal Reasoning: a Jurisprudential Preface to *Law and Economics*" by Péter Cserne, is devoted to consequentialism and the role of economic arguments in law. He also shows that the typical normative claims of law and economics based on economic efficiency can be interpreted as consequence-based arguments of a special kind. From a doctrinal perspective, one particular concern about consequence-based reasoning seems to be whether the consequentialism inherent in effectiveness and responsiveness inevitably corrupt law's non-instrumental commitments to doctrines and principles associated with values such as coherence and procedural justice. In fact, modern law has to fulfil conflicting normative expectations by combining doctrinal coherence with effectiveness and responsiveness as stressed by the regulatory perspective. In the analytical core of the paper, Cserne analyses the conceivability, feasibility and desirability of the judicial appreciation of general social consequences of legal decisions. Referring to the philosophical, jurisprudential and institutional dimensions of the issue, he argues that in a modern constitutional democracy, the scope of consequence-based judicial reasoning is limited mainly by the expertise of courts.

In the essay "Is the Rationality of Judicial Judgements Jeopardized by Cognitive Biases and Empathy?" by Klaus Mathis and Fabian Diriwächter, the question initially raised is to what extent cognitive biases compromise the rational basis of judges' judgements. As Amon Tversky and Daniel Kahnemann asserted in their essay "Judgement under Uncertainty: Heuristics and Biases" (1974), people in complex decision-making situations resort to what are known as mental heuristics. These cognitive rules of thumb are certainly useful, and even necessary, in order to cope with complex situations in everyday life, but sometimes lead to serious systematic errors, i.e. biases.

Building on these insights, Mathis and Diriwächter examine judges' decision-making, illustrating the problematique principally with reference to the assessment of negligence in tort law. They proceed to show how and to what extent cognitive errors can be avoided, or their impact at least mitigated or compensated. At the end of the essay, the related yet different question is raised as to whether a judge should have empathy, and whether this does not jeopardize judicial impartiality, and hence – like cognitive errors – ultimately, the rationality of judges' decision-making.

Part II, "Law and Sustainability", begins with Paolo Becchi's essay "Our Responsibility Towards Future Generations". He points out that Hans Jonas, in his foundational work *Das Prinzip Verantwortung (The Imperative of Responsibility)* published in 1979 (1984), was the first to recognize the nature and practical implications of technology, and to emphasize the necessity of a new ethics for the technological civilization. Jonas founds his ethics of responsibility on a metaphysics of humankind's ontological structure of being. Drawing from Aristotelian natural philosophy, he attempts to show that ends and values are not subjective attributions by humans but can be "derived from the being of the things themselves". In order to justify a categorical duty to preserve nature and humanity, he seems satisfied to demonstrate that purposiveness of the entirety of nature, culminating in the human being, has value as such and is worthy of preservation. He therefore attempts to demonstrate the existence of an objective, intrinsic purposiveness of nature. This purposiveness represents a "good as such", from which Jonas finally derives a categorical duty to conserve this good. The first imperative to be derived is, "*that* there be a mankind". Next, Jonas reformulates Kant's categorical imperative as the "imperative of responsibility": "Act so, that the effects of your action are compatible with the permanence of genuine human life."

Jonas was more than aware that he was tackling a highly contemporary subject with a decidedly less than contemporary, indeed almost archaic, philosophy. For he knew that his value-objectivism ran counter to two entrenched dogmas of our times, namely that there is no metaphysical truth and that an "ought" cannot be derived from an "is". In contrast to Max Weber, who justified his ethics of responsibility by means of the value-freedom of science, Jonas' concern here is specifically to overcome this value-freedom, which he believes to be the cause of the current predicament. In this step resides both the strength but also the weakness of Jonas' approach: anyone who rejects his ontological value basis will not find his argumentation of much use.

Following his appraisal of Hans Jonas, Becchi ventures to compare him with H. L. A. Hart, whose major work *The Concept of Law* (1961) included a chapter entitled "The Minimum Content of Natural Law" in which he formulated minimal conditions for a functioning legal system. Hart's assumption is that no moral or legal code can be put in place unless certain physical, mental or economic conditions are satisfied. In considering the survival of humankind as the primary purpose of law, Hart – according to Becchi – instils this content into the system of law. Becchi infers from this that Hart thus espouses the same teleological view of nature assumed by Jonas and, also like Jonas, directs attention towards survival as the proper end of human activity. This is a very original and unexpected comparison indeed. However, since Hart's argumentation is meant purely descriptively rather than normatively, these conclusions might be overdone.

Becchi, recognizing the obvious deficiencies of both approaches, concludes by asking, in connection with the ethics of the future, whether one should not reintroduce the category of the sacred, which has seemingly suffered irreparable damage under the prevailing technico-scientific approach. Here he seeks refuge in human dignity, which he derives in Christian tradition: God has not bequeathed us his creator role, Becchi argues, but has created the human being "as his likeness" and thus bestowed on us that transcendental *dignitas* which conferred on us a special position within nature. Before and beyond that value, he speculates, perhaps reference to the specificity of the *conditio humana* and its ontological meaning might allow us to slow down or even halt the present lunatic trajectory of the biotechnological society towards self-destruction. Perhaps, he speculates, this is also an anchor of hope for future generations: for we would have no right to rob them of that *dignitas* which characterizes us, and consign the human race to becoming a relic in evolutionary history.

The next essay, "Future Generations in John Rawls' Theory of Justice" by Mathis, is also devoted to the theme of intergenerational justice. In *A Theory of Justice* (1971), one of the subjects that Rawls dealt with was justice between the generations. A contract-theory justification of intergenerational justice is confronted by the following question: why should existing generations take on any obligations towards future generations? Future generations, unlike people living contemporaneously, lack any potential to pose a threat, which according to the classic contract theory is what drives individuals, out of rational self-interest, to adopt an ordered society in place of the unregulated state of nature. Therefore the main problem is rooted in the non-simultaneity and the consequent lack of reciprocity between present and future generations. In order to expand his theory to future generations, Rawls reflects on the original position from the perspective of multiple generations, and discusses intergenerational justice under the aspect of the "just savings" principle, whereby capital accumulation serves the purpose of establishing and maintaining a just system over time.

However, Rawls rejects the conceptualization of the original position as a gathering of all generations, because he believes that it would stretch the imagination too far. Instead, the parties know that they belong to the same generation, but not which one. If the generation represented in the original position is selfish, it has no reason to save for future generations, because it gains no benefit itself by doing so. To ensure that the parties in the original position nevertheless agree on a just savings principle, Rawls declares them to be representatives of ancestral lines whose immediate progeny matter to them. Later he gives up this rather unconvincing solution and proposes that the parties in the original position should choose the savings principle that they would wish past generations to have followed and future generations also to follow.

Since Rawls addresses the question of intergenerational justice only from the viewpoint of just saving, and does not treat natural capital as a special category, he makes a tacit assumption that the various types of capital are substitutable. This implies the concept of weak sustainability, which barely does justice to the ecological aspect of sustainability. This crucial theme is taken up once again in Mathis' essay, "Discounting the Future?" in connection with discounting as an element of cost-benefit analysis.

The next essay, Malte-Christian Gruber's "What Is It Like to Be Unborn?", continues the discussion of the importance of recognizing future generations. It outlines the problems raised by the inability to properly take into account the interests of these, as yet unborn, generations when making decisions. Problem areas include the inability to compensate future generations for damages inflicted by current generations on the environment which will affect them, too, such as the environmental damage as a result of oil spills. Gruber substantiates the link between current generations and future, as yet unborn generations through the human genome, which he defines as part of our common heritage.

Many problems have to be faced in this endeavour to give due recognition to future generations. These include the potential trade-off in efficiency, since efficiency only takes into account the interests of the current generations, the inability to assess with accuracy what the actual interests of future generations truly are, and the fact that it is impossible to eliminate every possible risk of irreparable damage to the environment without running the risk of preventing innovation. By engaging with the fiction of future generations as communities of fate, law gains the capacity to reach out beyond the boundaries of its own conceptual limitations. Biodiversity can therefore never be treated solely as a resource for economic exploitation, according to Gruber, since it also has major significance for physical survival in the future.

While Gruber sees biodiversity as an important aspect of sustainability, Milena Petters Melo teases out the importance of cultural heritage in her essay, "Cultural Heritage Preservation and Socio-environmental Sustainability: Sustainable Development, Human Rights and Citizenship". First, she discusses the history of the term "development" which has now found its way into the UN Declaration on the Right to Development (1986) as a human right in itself. Melo presents the various applications of the term "development" and how these have arisen over time. Like development, "cultural heritage" has also increased in importance. Migration has brought with it a series of difficulties; as a result, cultural heritage has been identified as playing a significant role for individuals with respect to intrinsic identity, appreciation of human rights and integration. Cultural heritage is described in this paper as not only referring to monuments in their tangible form but also to the intangible aspects that they represent.

The preservation of our cultural heritage is important as it helps to build solidarity on a global level as well as across generations and even future generations. A fallacy of sustainable development, according to Melo, is that the mainstream interpretation is not in support of protecting diversity of both nature and culture but is conceived more as a strategy for maintaining a continuing level of development. The focus should, however, be shifted to the protection of our cultural heritage in all its different forms, since cultural diversity, for Melo, is as necessary for humankind as biodiversity is for nature.

In Part III, "Law, Economics and Sustainability", the demands for efficiency and justice are brought together. This is brought into particular focus in Mathis' essay, "Discounting the Future?". For most economists, discounting in cost-benefit analysis is taken for granted and rarely questioned. On the other hand, philosophers largely seem to agree that it is unacceptable to place a lower value on future benefits and costs by means of discounting, since it breaches the precept of impartiality. After a detailed presentation of the complex debate, Mathis comes to the critical realization that cost-benefit analysis depends on the implicit assumption that all goods and types of capital are substitutable. Since this implies that natural capital is also substitutable, the question of discounting flows into the debate on "weak versus strong sustainability", i.e. is it possible and permissible to treat natural capital as interchangeable with other types of capital? The contentious dispute between philosophers and economists over discounting is therefore missing the real problem. Instead, Mathis urges that public projects should be subject to sustainability assessment in addition to cost-benefit analysis.

In his essay, "A Critical Review of 'Efficiency Ethics': The Case of Climate Economics", Felix Ekardt delivers a critique of the economic assumptions underlying current climate policy, and argues for a refinement of climate economics that moves away from conventional efficiency-based thinking and, instead, towards a general "climate social science". The starting point of his critique is the problem of the unsatisfactory status of the data used in connection with climate issues, which he attributes to the high degree of prognosis-uncertainty regarding the consequential impacts of climate change. These seemingly clear-cut economic figures conceal hidden assumptions and judgements, he contends. This ultimately leads to an unrealistic and unduly optimistic assessment of the development of climate change, and an underestimation of the potential extent of the resulting damage. This overoptimistic assessment by economists brings Ekardt back to the welfare economists' basic assumption that growth is unlimited, a perspective which he condemns as too narrow. It is time to abandon the neo-classical idea of endless economic growth, he urges, and pave the way for a paradigm shift.

In addition to these unacceptable empirical assumptions, however, he also points out that conventional climate economics is based on questionable normative assumptions. For instance, the preference and efficiency theory favoured by economists as a theoretical basis for assessing the correct climate policy is held to be inadequate because, for one thing, it fails to take account of the needs of future generations. Furthermore, he argues, the necessary monetary assessment of preferences does not work, because the latter simply cannot be quantified and expressed in terms of prices. Ekardt opposes this normative preference theory with a rational discourse-ethical theory of justice, the fundamental principle of which is freedom of the individual. Only by recognizing the autonomy and equality of every individual, he claims, is it possible in any debate on climate protection to take future generations into account as potential discourse participants, and in this way to find a solution that is intergenerationally equitable.

In comparison to Ekardt, the position taken by Balz Hammer in his essay "Valuing the Invaluable?" is less hostile towards the use of economic analysis methods for legal practice. The focal question of his analysis is whether, and how, a human life can be valued for purposes of assessing the cost-efficiency of state regulatory projects to mitigate mortality risks – e.g. traffic safety precautions – and whether this is acceptable at all from moral and legal viewpoints. Whereas economists essentially value all goods in units of money, the opinion is widespread among philosophers and lawyers that this is unacceptable because a human life has infinite value. One of this author's central concerns is therefore to intermediate in this dispute between economists and philosophers, and to draw out the practical implications for state decision-makers.

Hammer begins by setting out the different methods of economic analysis for the valuation of a human life, and subjects them to a critical methodological review. He goes on to assess the valuation of human life from a legal and ethical perspective. The primary question at issue is whether a human life – as some claim – is of infinite value. In the course of his attempt to bring clarity to this question, the importance of differentiating between descriptive and normative statements is underscored once again. According to Hammer, this important distinction not only helps to reconcile the conflict between economists and philosophers on the theoretical level, but is also shown to be fundamental for regulatory policy insofar as it relies on economic calculations to assess the cost-efficiency of risk-mitigating projects.

The ten essays stand as a collection of great methodological and thematic diversity. Common to all of them, however, are interdisciplinary inquiries at the interfaces of law, economics and philosophy. The different academic disciplines are increasingly separating out into different directions and generating knowledge which is only appreciated – if at all – intersubjectively within the discipline as meaningful and problem-relevant. This fosters a strong temptation to forget that reality is adisciplinary and that real problems can usually only be solved by combining knowledge from different cross-disciplinary areas. The great value of interdisciplinary studies is their capacity to nurture dialogue between the disciplines and, as a result, to clear up all manner of misunderstandings. Like Economic Analysis of Law, sustainability is very much a typical cross-cutting theme which cannot easily be assigned to a single discipline. The theme of sustainability not only brings together different academic disciplines but, mediated by the interest in intergenerational justice, opens up the horizon of vision beyond the present into the future.

Contents

Contributors

Paolo Becchi University of Lucerne, CH-6002 Lucerne, Switzerland; University of Genoa, Liguria, Italy, paolo.becchi@unilu.ch; paolo.becchi@unige.it

Péter Cserne Tilburg Law School, Tilburg University, NL-5000 LE Tilburg, The Netherlands, peter.cserne@uvt.nl

Fabian Diriwächter Lucerne, Switzerland, fabian.diriwaechter@stud.unilu.ch

Felix Ekardt Baltic Sea Institute for Environmental Law, Research Group Sustainability and Climate Policy, Könneritzstrasse 41, D-04229 Leipzig, felix.ekardt@uni-rostock.de

Malte-Christian Gruber Institute of Economic Law, Goethe University Frankfurt, Grüneburgplatz 1, D-60629 Frankfurt am Main, Germany, gruber@jur.uni-frankfurt.de

Balz Hammer University of Lucerne, Faculty of Law, Frohburgstrasse 3, P.O. Box 4466, CH-6002 Lucerne, Switzerland, balz.hammer@unilu.ch

Klaus Mathis University of Lucerne, Faculty of Law, Frohburgstrasse 3, P.O. Box 4466, CH-6002 Lucerne, Switzerland, klaus.mathis@unilu.ch

Milena Petters Melo Research Centre on European Institutions, University of Naples Suor Orsola Benincasa, Naples, Italy; Environmental Constitutional Law and Comparative Studies, University of Blumenau – FURB, Blumenau, Brazil; University of Curitiba – UNIBRASIL, Curitiba, Brazil; University of Salento, Lecce, Italy, pettersmelo@libero.it

About the Authors

Paolo Becchi, Lucerne and Genoa. Professor of Philosophy of Law at the University of Lucerne and University of Genoa. CH-6002 Lucerne, Frohburgstrasse 3, P.O. Box 4466, Tel. + 41 (0)41 229 53 87; Fax + 41 (0)41 229 53 01. paolo.becchi@unilu.ch; paolo.becchi@unige.it. Fields of interest: Legal Philosophy, in particular the Philosophies of Hegel and Hans Jonas, the Age of Enlightenment and the Codification of Law; also Bioethical Questions (Medically Assisted Insemination, Cloning, Euthanasia, Brain-Death and Organ-Transplantation).

Péter Cserne, Tilburg. Assistant Professor at Tilburg Law School and Senior Research Fellow of Tilburg Law and Economics Centre. NL-5000 LE Tilburg, Warandelaan 2, P.O. Box 90153, Tel. + 31 (0)13 466 36 45, Fax + 31 (0)13 466 80 47. peter.cserne@uvt.nl. Fields of interest: Economic Analysis of Law, Legal and Political Theory, Comparative Private Law, Contract Law and Theory.

Fabian Diriwächter, MLaw, Lucerne. fabian.diriwaechter@stud.unilu.ch. Fields of interest: Competition Law, Civil Law, Economic Analysis of Law.

Felix Ekardt, LL.M., M.A., Leipzig and Rostock. Professor of Environmental Law and Legal Philosophy, Baltic Sea Institute for Environmental Law, Research Group Sustainability and Climate Policy. D-04229 Leipzig, Könneritzstraße 41, Tel./Fax + 49 (0)341 926 08 83. felix.ekardt@uni-rostock.de. Fields of interest: Environmental Law, Climate Law and Climate Policy, Theory of Human Rights, Transdisciplinary Theory of Sustainability, Theory of Justice, WTO Law.

Malte-Christian Gruber, Frankfurt am Main. Assistant Professor (Akademischer Rat) at the Institute of Economic Law, Goethe University Frankfurt. D-60629 Frankfurt am Main, Grüneburgplatz 1, P.O. Box 31, Tel. + 49 (0)69 798 34254; Fax + 49 (0)69 798 763 34254. gruber@jur.uni-frankfurt.de. Fields of interest: Private Law (Civil Law, Information Technology Law, Biotechnology Law), Bioethics, Legal Philosophy and Legal Theory.

Balz Hammer, MLaw, Lucerne. Researcher and Ph.D. student at the Chair for Public Law and Law of the Sustainable Economy of Prof. Dr. Klaus Mathis at the University of Lucerne. CH-6002 Lucerne, Frohburgstrasse 3, P.O. Box 4466, Tel. + 41 (0)41 229 54 18; Fax + 41 (0)41 229 53 97. balz.hammer@unilu.ch. Fields of interest: Constitutional Law, Legal Philosophy and Economic Analysis of Law.

Klaus Mathis, Lucerne. Professor of Public Law and Law of the Sustainable Economy at the University of Lucerne. CH-6002 Lucerne, Frohburgstrasse 3, P.O. Box 4466, Tel. + 41 (0)41 229 53 80; Fax + 41 (0)41 229 53 97. klaus.mathis@unilu.ch. Fields of interest: Public Law (Law of Sustainable Development, Economic Constitutional Law), Economic Analysis of Law, Legal Philosophy and Legal Theory.

Milena Petters Melo, Naples. Researcher at the Research Centre on European Institutions – University of Naples Suor Orsola Benincasa, Invited Professor of Environmental Constitutional Law and Comparative Studies at the University of Blumenau – FURB, Brazil, Invited Professor of Constitutional Law at the University of Curitiba – UNIBRASIL, Coordinator of the Euro-American Didactic Centre on Constitutional Policies – University of Salento, Italy. I-80122 Napoli, Salita Piedigrotta, n° 3 (c/o Mancusi-Fabricatore), Tel. + 39 3291 528 861; + 55 48 99 108 937. pettersmelo@libero.it. Fields of interest: Constitutional Law, Comparative Law, Environmental Law, International Law (Human Rights and International Cooperation), Legal Philosophy and Legal Theory, Humanities and Cultural Studies.

Part I
Law and Economics

Consequentialism in Law

Klaus Mathis

> *The proof of the pudding is in the eating, not in the cookery book.*
> *–Aldous Huxley*

This essay analyses the significance of consequentialism in legislation and legal adjudication. After a short discussion of legislative impact assessment, the debate on consequentialism in legal adjudication is presented in detail, making particular reference to the situation in Germany, Switzerland and Anglo-American countries. By way of exemplification, the discussion moves on to consider the application of the Hand rule in tort liability.

1 Introduction

In recent times, the law has appropriated consequentialism from two principal standpoints: that of economics, via economic analysis of law, and that of ecology, via environmental impact assessment and technology impact assessment.[1] This essay focuses primarily, but not exclusively, on the application of economic analysis concepts in law.

Whereas the principle of impact analysis during the regulatory process is generally undisputed, opinions are divided over the question of whether to allow consequential arguments to carry any weight in legal adjudication. If we accept that the application of law may involve an element of creating new law, and that the courts[2] thus assume a legislative function in certain areas, there is no avoiding the question of whether, in that case, they should not also consider the consequences of their decision during the deliberation process.

---

[1]Deckert, *Folgenorientierung*, p. 1.

[2]This also applies to the administrative authorities in a similar way.

K. Mathis (✉)
University of Lucerne, Faculty of Law, Frohburgstrasse 3, P.O. Box 4466, CH-6002 Lucerne, Switzerland
e-mail: klaus.mathis@unilu.ch

K. Mathis (ed.), *Efficiency, Sustainability, and Justice to Future Generations*, Law and Philosophy Library 98, DOI 10.1007/978-94-007-1869-2_1,

2 Consequentialism in the Regulatory Process

The instrument of legislative impact assessment is in use in the majority of countries today, albeit in varied ways. Since 1995, the OECD has recommended to its member countries to carry out a Regulatory Impact Analysis (RIA) as part of the legislative process. A priority OECD initiative towards that end is SIGMA (Support for Improvement in Governance and Management in Central and Eastern European Countries).[3] In the European Union, in the wake of the Mandelkern Report (2001) a plan was adopted to simplify and improve the regulatory framework. It prescribes an impact assessment for the most important legislative proposals.[4] In Germany, the procedure is set out in the "*Leitfaden zur Gesetzesfolgenabschätzung*", a guide to legislative impact assessment commissioned by the German federal government and published in July 2000.[5] In Austria, a reference manual entitled "*Was kostet ein Gesetz?*" ("What is the cost of a law?") was published in 1992, although it is rather business-administrative in thrust.[6] It is also becoming increasingly common for legislative impact assessment to apply the tools of economic analysis. In the United States, for example, cost-benefit analyses have long been standard practice in relation to major new regulations.

Since the year 2000, Switzerland has had its own instrument of regulatory impact assessment (*Regulierungsfolgenabschätzung*, RFA) at the Confederation level, which is informed by the OECD recommendations.[7] Its constitutional basis is given in Art. 170 of the Swiss Federal Constitution, according to which the Federal Assembly (the Swiss parliament) has to ensure scrutiny of the effectiveness of government measures.[8] The specific statutory hook for undertaking a prospective analysis of draft legislation is found in Art. 141 para. 2 lit. g of the Swiss Parliament Act. Under this provision, the notices to draft bills proposed by the Swiss Federal Council must include statements of impacts on the economy, society and the environment, insofar as substantial comments on these aspects can be made.

According to the decree and guidelines of the Swiss Federal Council of 15 September 1999, all legislation must therefore be subjected to an economic impact analysis before it is enacted. The analysis should include scrutiny of the following five points: (1) Necessity and possibility of state action; (2) Impacts on individual social groups; (3) Impacts on the whole economy; (4) Alternative regulations; (5) Expediency in enforcement. So far, regulatory impact assessment has

[3]Weigel, pp. 194 ff.

[4]Bundesministerium des Innern, Der Mandelkern-Bericht – Auf dem Weg zu besseren Gesetzen. On this approach, see also Andrea Hanisch, Institutionenökonomische Ansätze in der Folgenabschätzung der Europäischen Kommission.

[5]Carl Böhret and Götz Konzendorf, Leitfaden zur Gesetzesfolgenabschätzung; see id., Handbuch der Gesetzesfolgenabschätzung (GFA), also Matthias Dietrich, Folgenabschätzung von Gesetzen in Deutschland und Großbritannien.

[6]Weigel, p. 195.

[7]See OECD, Regulatory Impact Analysis.

[8]On the situation in Switzerland, cf. also Mader, pp. 100 ff.

essentially been used in Switzerland as a prospective means of analysis in the context of finalizing legislation at Confederation level. Supplementary use is made of another instrument, the small- and medium-sized enterprise compatibility test (*KMU-Verträglichkeitstest*). For major regulations, a cost-benefit analysis is also required.

The strength of *cost-benefit analysis* is that it attempts a comprehensive evaluation of the economic impacts of a measure or a project. But mention must be made of this method's weaknesses, too: the insistence on monetarization means that a financial value must be attached to all impacts, even those for which no market prices are available. Whilst it is relatively easy to evaluate costs in monetary terms, benefits must often be assessed using ad hoc reference data and rough approximations. These uncertainties produce valuations with rather broad margins of error, which can cast doubt on the meaningfulness of the results. Moreover, future costs and benefits must be discounted to a reference point in time. Here the choice of the discount rate has significant implications for the result.[9]

A further aspect to bear in mind is that under cost-benefit analysis, it makes no fundamental difference which social groups will be the beneficiaries of a legal regulation and who will have to bear the likely costs. It is sufficient if society's balance sheet is positive after all costs and benefits have been accounted for. In the terminology of welfare economics, what this means is that only the Kaldor-Hicks criterion for compensation needs to be satisfied.[10] It would therefore be desirable if, in every case, legal regulations were always analysed with regard to their impact on income distribution as well, to enable political decision-makers to arrive at a conclusive overall judgement.

3 Consequentialism in the Application of Law

Before discussing the consequences of judges' decisions, it is important to be clear about which type of consequences are meant here. A key distinction can be made between *legal consequences* and *real consequences*. According to Lübbe-Wolff, legal consequences are those consequences attached by legal provisions when certain preconditions are met. Real consequences, in contrast, are the *actual* consequences of the validity and application of legal provisions.[11] Real consequences can be further subdivided into the consequences for the parties directly affected by the judgement (*micro-level real consequences*) and consequences for the whole of society (*macro-level real consequences*).[12] A similar distinction is made between the

[9]On the questions surrounding cost-benefit analyses, see e.g. Lave, 'Benefit-Cost Analysis'. See also in the present volume Klaus Mathis, 'Discounting the Future?' and Balz Hammer, 'Valuing the Invaluable?'.

[10]For further details, see Mathis, pp. 56 ff.

[11]Lübbe-Wolff, p. 25.

[12]Thus van Aaken, pp. 171 f.; cf. also Wälde, p. 6. Sambuc, pp. 101 ff. talks about individual consequences and social consequences. On the whole dabate, cf. also Deckert, Folgenorientierung, pp. 115 ff.

direct consequences arising from the events of the case, and the consequences of the judgement's precedent effect on the future behaviour of the immediate parties and of all other norm-addressees[13].[14]

A further consideration is that the methodology of impact analysis consists of a *positive* and a *normative* element: the first step is to identify the expected impacts, and the second step is to evaluate these impacts. Impacts are not good or bad per se; the same impacts can be evaluated positively or negatively depending on values or party status. Identification of the impacts calls upon *sociological, technological or psychological knowledge* in particular; for the evaluation of the impacts, *normative criteria* are required.

One difference between the traditional legal method and the economic method is whether the case is considered from a *retrospective* or a *prospective* point of view. From a lawyer's perspective, it is usual to assess a concrete case that has already occurred (the retrospective view). As far as the impacts of the judicial decision are concerned, the primary focus is then on the immediate micro-level real consequences. Macro-level real consequences are less commonly addressed.[15] From the economic perspective, however, the macro-level real consequences are of greatest interest, with specific regard to the judgement's precedent effect on the future behaviour of all addressees of the norm (the prospective view). The evaluation criterion in this case is *allocative efficiency.*

An example may serve to explain this: in tort liability, the jurist is concerned with settling a claim between the injurer and the injured party. The latter wants to be compensated for damage suffered as a result of a harmful incident that has occurred. From an economic perspective, the question is framed differently: what is the impact of the judgement on the future behaviour of potential injurers and injured parties, i.e. how will it affect their behaviour with regard to precautionary measures or activity levels, what costs and benefits arise and how does it change allocative efficiency or social welfare?

From a legal perspective, the fundamental question about impact analysis is which of the specified consequences are legally material for the judge; in other words, clarification is required as to which consequences the law is receptive to, and which ones may – or must – be taken into account in the application of law.[16] Of course, cases in which the law explicitly instructs the court to take certain consequences into account are unproblematic. In the other cases, the answer will hinge on whether taking decision-impacts into account is qualified as "*legal policy*", in

[13]In penal law, this is called special prevention and general prevention.

[14]Lübbe-Wolff, pp. 139 ff., similarly distinguishes between "decision-impacts" *(Entscheidungsfolgen)*, i.e. the impact on the parties when an authoritative judgement is pronounced (e.g. imprisonment, fines and their repercussions for convicted individuals and their families, etc.) and "adaptation-impacts" *(Adaptationsfolgen)*, i.e. the general influence of legal rules on behaviour.

[15]But there are certainly juristic concepts, such as the "public interest" or "proportionality", which presuppose the consideration of macro-societal interests or associated ends-means relations.

[16]Hoffmann-Riem, p. 38.

which case they would have no place in the "*legal*" reasoning of the courts, or whether they are deemed to be an element of the juristic programme of judgement and reasoning.[17]

Interestingly, this question is not only a prominent topic of debate in continental European law but in the Anglo-American world as well. The following essay will therefore outline both the discussion in the German-speaking literature and the equivalent controversy in the Anglo-American discourse.[18]

3.1 Arguments Against Considering Impacts

In the German-speaking world, Niklas Luhmann might be seen as the fiercest opponent of consequentialism in the application of law. Legal adjudication, in Luhmann's thinking, is *conditionally programmed* by the legislator. If certain conditions are fulfilled (facts amounting to breach of a legal provision) then a certain judgement has to be reached (the "if-then form").[19] In this way, highly complex matters can be resolved into congruently predictable judgements (stabilization of expectations).[20] Luhmann thus draws a line between programmed judgements (application of law) and programme-defining decisions (legislation). The apparent one-sidedness of conditional programmes can be corrected at higher decision-making levels by passing statutes and by modifying conditional programmes as a result of policy decisions made with particular goals in mind.[21]

Luhmann therefore rejects consequentialism in the application of law, reasoning that foisting socio-political consequentialism on a legal system runs the risk of compromising its dogmatic autonomy and disorientating it completely, turning it away from criteria that transcend the decision-making programme, indeed from any criteria except the consequences themselves.[22] He therefore argues in favour of relieving the application of law of the burden of responsibility for consequences:

> Such relief from full responsibility for consequences is necessary if the conditional programme's function, namely to give a reliable expectation of future sequences of events, is to be fulfilled. The principle of equality likewise cements this structure. Were the jurist to be made responsible for situation-specific consequences of his judgement, he would have to absorb and process completely different information in the adjudication process, he would have to develop a completely different style of work and examination, carry out forecasts,

[17]Neumann, 'Theorie', p. 234.

[18]The literature on the consequences problem has become enormously vast, and for this reason the following presentation makes no claim to exhaustiveness.

[19]Luhmann, *Rechtssoziologie*, p. 227.

[20]Luhmann, *Rechtssoziologie*, p. 229. Id., 'Argumentation', p. 29, sees the role of juristic argumentation as the means whereby the system reduces the individual's experience of surprise to a tolerable level.

[21]Luhmann, *Rechtssoziologie*, p. 234. On implementation of the conditional programming in the administration, see id., *Automation*, pp. 35 ff.

[22]Luhmann, *Rechtssystem*, p. 48.

> probability analyses, cost-effectiveness calculations and assessments of side-effects. This would render him an unpredictable element, and all the more so the more rationally he proceeded.[23]

Accordingly Luhmann advances three arguments against consequentialism in the application of law: the argument of *legal certainty*, the argument of *legal equality* and the argument of *overburdening the courts*. A fourth argument cited elsewhere is that consequentialism jeopardizes the *independence of the courts.*[24] For Luhmann, it is absolutely impossible to see how legal questions can be cross-referenced with sociological theories or the methods of empirical social research in the degree of detail necessary to reach a decision.[25]

Luhmann's critique suffers from a reliance on assertions or assumptions that are only set forth on an abstract level. It seems as if he has barely, if ever, given serious thought to how the consequentialism that he so vehemently criticizes might be envisaged in practice.[26] Quite clearly, he is labouring under the misapprehension of impact analysis as case-specific, unsystematic and randomistic. He fails to recognize that even for the purpose of an impact analysis, a case must not be treated as a unique event but must be seen as an exemplar of an entire genre of similar cases.[27] *Even impact analysis has to adhere to a defined schema and logic and should be channelled accordingly by means of dogmatic structures.* Moreover, it must always operate within the scope permitted by statutory definitions and legal consequences.[28] Accordingly, even when impact arguments are taken into account, a *constant and coherent ruling practice* will become settled. As long as consequentialism is kept within the bounds of these constraints, neither legal certainty nor legal equality nor the independence of the courts are in any jeopardy.

Only the argument of the possible *overburdening of the courts* represents a genuine problem. The problem is manifested not only on the cognitive but also on the normative level:[29] the first step is to identify the consequences, a task which in some instances may demand considerable sociological knowledge and involve substantial effort. Further factors to consider are that the courts are compelled to give judgements, and process economy dictates that cases should be settled within a useful time-frame. But even when all the consequences are known, they still have to be evaluated in the second step. Some consequences may be patently desirable or undesirable. Frequently, however, consequences arise which prove more debatable or which have positive impacts on certain parties and negative impacts on others, so that the net effect is unclear. Nevertheless, it is irrational to demand the complete abandonment of impact analysis because of these practical problems.

---

[23] Luhmann, 'Methode', p. 4 (own trans.). Likewise id., *Ausdifferenzierung*, p. 276.

[24] Luhmann, *Rechtssoziologie*, p. 232. See also id., *Ausdifferenzierung*, pp. 140 ff. and pp. 275 ff.

[25] Luhmann, *Rechtssystem*, p. 9.

[26] Koch and Rüßmann, p. 234.

[27] Koch and Rüßmann, p. 234.

[28] Rottleuthner, pp. 114 f.

[29] Seiler, *Rechtsanwendung*, pp. 57 ff.

Rather, the aim must be to carry out the impact analysis as far as possible.[30] Taking account of sociological knowledge undoubtedly represents progress compared to the often-criticized commonsense theories of judges.[31]

Esser finds Luhmann's idea of programmed application of law not only wrong but highly dangerous:

> The total systemic autonomy of the law, on the other hand, prohibits any critical reflection by the interpreter upon the conditions and motivations of his predisposition and thus also prevents any rational verification and delivers the legal system, for all its ideological insulation and autonomous comportment, into the arms of political manipulation.[32]

Another critical voice is Rhinow's, who contends that Luhmann's view is based on *outdated judicial ideas* and that it obscures rather than enriches the legal theoretical debate;[33] the law to be applied is not in fact a gapless, ready-made "programme", he argues, and therefore the concretization process cannot be interpreted primarily as programmed judgement, because those responsible for applying the law are themselves also producers of valid law.[34]

Despite a longstanding tradition of consequentialism in Anglo-American law,[35] eminent critics have been equally outspoken on the matter of consequentialism in the application of law. One is Ronald Dworkin, who in "Hard Cases" (1975)[36] draws a distinction between "principles" and "policies" (the latter being arguments in the interests of the common good or collective goals) and calls for courts to confine themselves strictly to arguments of principle, even when ruling on hard cases, because arguments based on policy goals are the preserve of the legislator.[37] According to Dworkin, rights can be derived from principles whereas arguments based on policy goals cannot:[38]

---

[30]Koch and Rüßmann, p. 235. Approaches to this are found in Martina R. Deckert, 'Praktische Durchführbarkeit folgenorientierter Rechtsanwendung – Auf dem Weg zu einer folgenorientierten Rechtswissenschaft'.

[31]Cf. Rottleuthner, pp. 115 f.

[32]Esser, *Vorverständnis*, p. 141, see also pp. 205 ff. On the concept of predisposition, see Section 3.2, below.

[33]Indeed, Luhmann does appear to base his view on Montesquieu's long-superseded idea that the judges are only a subsumption machine or "la bouche qui prononce les paroles de la loi" (Montesquieu, XI, 6). On this, see also Regina Ogorek, *Richterkönig oder Subsumtionsautomat? Zur Justiztheorie im 19. Jahrhundert.* Grimm, pp. 140 f., also points out that in addition to classic conditional norms, it is increasingly common – particularly in public law – to encounter usually vaguely-formulated *final norms* (goal specifications, policy programmes). This is quite plainly another circumstance that has reinforced the necessity of the consequentialist view in legal dogmatics (cf. Sommermann, p. 53). See also Klaus Hopt, 'Finale Regelungen, Experiment und Datenverarbeitung in Recht und Gesetzgebung'.

[34]Rhinow, p. 256.

[35]On which, see Section 3.2 below.

[36]Dworkin, *TRS*, pp. 81 ff.

[37]Dworkin, *TRS*, p. 85; Watkins-Bienz, p. 83. See also Bittner, pp. 227 ff.; Wolf, pp. 364 ff.; Harris, pp. 201 ff.

[38]Harris, p. 201.

> Arguments of policy justify a political decision by showing that the decision advances or protects some collective goal of the community as a whole. The argument in favor of a subsidy for aircraft manufacturers, that the subsidy will protect national defense, is an argument of policy. Arguments of principle justify a political decision by showing that the decision respects or secures some individual or group right. The argument in favor of anti-discrimination statutes, that a minority has a right to equal respect and concern, is an argument of principle.[39]

Judges are always bound to the law in determining the content of legal norms, according to Dworkin, and must not create new law themselves; nor do they have any discretion.[40] According to his "rights thesis" the law can be conceived of as a system of individual rights, which are to be enforced with the help of judicial rulings in favour of one party or another, and within which one party always has the right to win the legal dispute, even in hard cases.[41] Linked to this is the "one right answer thesis" which claims that even in hard cases, the courts can and must deliver the sole and only right judgement.[42]

Dworkin demonstrates how the judge's role should be envisaged within his model by invoking Hercules, a fictive judge endowed with superhuman intellectual abilities;[43] the one and only right answer exists even in hard cases, he asserts, and can be identified by a judge endowed with Herculean abilities. If a judge believes the law to be incomplete, incoherent or imprecise, then the limits of his intellectual capacity are to blame and not any putative imperfection of the law.[44] H. L. A. Hart sums up Dworkin's theory and its implications for adjudication superbly as follows:

> [A] judge who thus steps into the areas of what he calls policy, as distinct from principles determining individual rights, is treading forbidden ground reserved for the elected legislature. This is so because for him not only is the law a gapless system, but it is a gapless system of rights or entitlements, determining what people are entitled to have as a matter of distributive justice, not what they should have because it is to the public advantage that they should have it.[45]

In justification of this rejection of policy arguments in legal adjudication, the principal arguments put forward by Dworkin are democracy, retroactivity and coherence. According to the *democracy argument*, judges are not generally elected by the people, or at least are not answerable to the electorate in the same way as the legislature, and consequently they are not legitimized to make new law.[46] In the *retroactivity argument*, Dworkin points out that it would be an outrage if a party lost

---

[39] Dworkin, *TRS*, p. 82.

[40] Greenawalt, 'Discretion', p. 361.

[41] Dworkin, *TRS*, pp. 81 ff.; Watkins-Bienz, p. 83.

[42] Dworkin, *TRS*, pp. 279 ff., and id., *MP*, pp. 119 ff. A critique of the obviously dubious one-right-answer thesis is superfluous here. See e.g. Ott, pp. 183 ff.; Watkins-Bienz, pp. 117 ff.; Bittner, pp. 233 ff.; Auer, pp. 85 ff.

[43] Dworkin, *TRS*, pp. 105 ff., and id., *LE*, pp. 239 ff.

[44] Hart, p. 183.

[45] Hart, p. 141.

[46] Dworkin, *TRS*, p. 84.

an action for breaching a duty that had only been imposed upon them *ex post facto*.[47] Finally, Dworkin fears that the admissibility of policy arguments would result in a *loss of coherence* in adjudication. Application of law in the case of principles arguments must be in harmony with previous cases, which guarantees equal treatment, he argues, whereas the achievement of particular policy goals does not necessarily require any such equal treatment.[48]

In Dworkin's model, a court that practices coherent application of the principles and concludes that a plaintiff has a claim for damages for negligent medical treatment is not allowed to be influenced by the policy argument that such actions drive up the costs of precautionary measures in the healthcare system to exorbitant levels.[49] Hence, the court – and this is the significant conclusion for the question of consequentialism – has to decide solely on who is entitled to what rights, but must refrain from considering the real consequences of its judgement for the parties affected and for society.

Hart's criticism of Dworkin's model is that a judge ruling on a new case must seek a general principle that could explain both the rulings on past cases of the same kind as well as the new case. He contends that a variety of principles are quite likely to be found which fit the past rulings but which would deliver different solutions to the new case.[50] Consequently, determining the right answer is not just a problem of knowledge. Even a judge equipped with superhuman qualities like Hercules could not determine the one right answer in these circumstances. Let us also recall Kelsen's even earlier warning that the fiction of the single correct interpretation of a legal norm should be avoided:

> Jurisprudential interpretation must take the utmost care to avoid the fiction that a legal norm only ever admits of one interpretation, which is the 'correct' one. That is a fiction used by traditional jurisprudence to uphold the ideal of legal certainty. In view of the ambiguity of most legal norms, this ideal can only approximately be realized.[51]

Finally, Hart points out that Dworkin's staunch rejection of any consideration of societal consequences during legal adjudication would conflict with the views held by many jurists, and was just an expression of Dworkin's hostility to utilitarianism:

> This exclusion of 'policy considerations' will, I think, again run counter to the convictions of many lawyers that it is perfectly proper and indeed at times necessary for judges to take account of the impact of their decisions on the general community welfare. [...] Professor Dworkin's exclusion of such considerations from the judge's purview is part of the general hostility to utilitarianism that characterizes his work [...].[52]

---

[47] Dworkin, *TRS*, p. 84.

[48] Dworkin, *TRS*, p. 88.

[49] Harris, p. 202.

[50] Hart, p. 139.

[51] Kelsen, *RR*, p. 353 (own trans.).

[52] Hart, p. 141. On Dworkin's reply to Hart's criticism, see the appendix in Dworkin, *TRS*, pp. 292 ff.

Neil MacCormick is another who disagrees with Dworkin's strict separation of principles and policy goals, seeing them not as opponents but, on the contrary, interrelating forces:

> [T]he spheres of principle and of policy are not distinct and mutually opposed, but irretrievably interlocking [. . .]. To articulate the desirability of some general policy-goal is to state a principle. To state a principle is to frame a possible policy-goal.[53]

According to MacCormick, arguments of rightness and the pursuit of goals are two sides of the same coin because the values of the legal order are an expression of the prevailing policy.[54] Likewise, Greenawalt comes to the conclusion that the distinction between principles and policy goals is blurred when every policy goal can be transformed into a legal principle and vice versa.[55]

To sum up, the main arguments that can be advanced against impact analysis in the application of law are the courts' *lack of legitimation and functionality* and the *risks to legal security and legal equality.*[56] The widespread hostility to considering impacts among lawyers may also stem from an *unholy alliance between conservative and progressive jurists.* The former fear that consequentialism, acting as a Trojan horse, might smuggle extra-legal and policy arguments into the citadel of juristic reasoning, and thereby soften up traditional dogmatics. The latter see consequentialism as opening the floodgates for utilitarian arguments, which could be used to justify sacrificing individual rights for the sake of pursuing societal goals, particularly when backed by cost-arguments.

3.2 Arguments in Favour of Considering Impacts

Auer points out that it is not just practically but also theoretically impossible to delimit "legal" from "legal policy" arguments (i.e. from consequentialist ethical and sociological arguments beyond the narrow bounds of statutory and intralegal assessments). Hence, no class of legally binding arguments can be isolated from the general societal discourse,[57] she claims, quoting support from Hans Kelsen and Josef Esser, among others.

Kelsen commented on this issue in the essay "*Wer soll der Hüter der Verfassung sein?*" (Who shall be the guardian of the constitution?; 1931), as follows:

---

[53] MacCormick, *Legal Reasoning*, pp. 263 f.

[54] MacCormick, 'Legal Decisions', p. 257. Similarly, Hiebaum, p. 86, who asserts that "every principles argument owes its persuasiveness to some, albeit unquestioned, 'policy orientation'." (own trans.).

[55] Greenawalt, 'Policy', pp. 1013 f. On Dworkin's reply to Greenawalt's criticism, see the appendix in Dworkin, TRS, pp. 294 ff.

[56] For a synopsis, see also e.g. Deckert, *Folgenorientierung*, pp. 13 ff.; Eidenmüller, pp. 414 ff.; Koch and Rüßmann, p. 227.

[57] Auer, p. 81.

> The opinion that only legislation but not the ‘real’ administration of justice is political is just as wrong as the opinion that only legislation is productive law-creation, whereas the judicial process is only reproductive application of law. [...] By empowering the judge within certain boundaries to weigh opposing interests and rule in favour of one or the other, the legislator confers on him an authority to create law, which is a power that gives the judicial function the same ‘political’ character, though not to such a high extent, as the legislator’s. Between the political character of legislation and that of the administration of justice there is only a quantitative, not a qualitative difference.[58]

Esser points out the issue of “predisposition” in the judicial interpretation of legal norms.[59] He is not referring to the personal social experience in which a judge grew up – and must also account for, of course – but means “predisposition” in the hermeneutic sense of some unavoidable anticipation regarding the question of meaning and outcome which is characteristic of a practical science.[60] Consequently he considers the idea of a value-neutral dogmatics to be naïve:

> It has a coarsening and distorting effect if one seeks to attribute these elements of [the judge’s] convictions to the categories of ‘legal policy’ and ‘legal dogmatic’ truth. Such key-words conceal the connectedness and the interplay of these decision-elements in the forming of convictions. [...] One is thereby failing to recognize that, as a matter of principle, in every decision ‘according to positive law’ both forces of legal consciousness exert an influence on the adjudication process: The striving for a ‘true-to-life’ and ‘reasonable’ decision in the light of the practical constraints and the demand of justice for legitimate expectations and conscientious responsibility for the preservation of a legal system, the stability of which only permits development in ‘small steps’.[61]

Further, Podlech and Sambuc also express support for impact analysis up to a point. Podlech shares Esser’s view that value-free application of law is non-existent or barely exists. From this he infers that for any assessment, a discussion must take place of the societal consequences of that assessment in terms of its rightness or wrongness and its acceptability to society. What he hopes for is a gain in rationality through this procedure, because it ensures that the arena of legal debate on the problem of value judgements is not left solely to irrational positions.[62] Sambuc argues that norms developed in the judicial process can be legitimized through the quality of their outcomes and/or through justification of their objectives. In this regard, the consideration of impacts could contribute to the quality of judicial rule-making by enabling a goal-instrumental pursuit of regulatory objectives, although without providing sufficient justification per se for the regulatory goals, i.e. the value preferences. Nevertheless, in his view, consideration of impacts could contribute to the justification of regulatory goals since they would be a means of determining whether

---

[58] Kelsen, ‘Verfassung’, p. 67. Here he is referring particularly, though not exclusively, to the constitutional judicial process.

[59] Esser, *Vorverständnis*, pp. 136 ff.

[60] Esser, ‘Möglichkeiten’, pp. 101 f.

[61] Esser, *Vorverständnis*, pp. 151 f. (own trans.).

[62] Podlech, p. 209.

prospective decision-impacts were compatible or incompatible with values and legal principles already positivized in law.[63]

Deckert points out in her study that widespread dissatisfaction with the merits and options of the conventional legal doctrine of methods is by no means insignificant as a rationale for consequentialism. The classical *canones* were in many cases inadequate to the task of providing orientation. The greatest dissatisfaction centres on the fact that observance of the individual interpretation criteria leads to different outcomes, and that no binding hierarchy exists according to the prevailing view.[64] She suspects that the standard practice amounts to choosing the interpretative outcome that leads to a satisfactory result on a case-by-case basis.[65]

Rüthers places particular emphasis on the unavoidability of considering impacts in the scope of *judicial development of law*,[66] because this inevitably means rule-making, and hence, engaging in legal policy. This would have two significant consequences: on the one hand, purely logical thought geared towards conceptual classification and differentiation is not sufficient to fulfil the legal policy function of dogmatics in the development of law. Dogmatics at its core is always influenced by values and worldview, according to Rüthers.[67] On the other hand, *active engagement in legal policy* also means *responsibility for impacts* and implies a requirement to weigh the consequences when putting forward dogmatic concepts and principles. Dogmatic statements are also instruments for shaping reality. For this reason, the predictable consequences of the shaping process cannot be disregarded and should be examined with reference to social interdependencies and findings from the economic and social sciences.[68]

Neumann rightly talks about the *Janus-headed nature* of legal reasoning: on the one hand this is highly authority-based (bound by the authority of the law, but also by the significance of precedents and the "prevailing opinion"). On the other hand, reasoning is becoming increasingly consequentialist, i.e. with reference to the expected impacts of the judgement.[69] Neumann demands that general impact considerations which would be of import to the judgement should be made transparent. This demand for honesty of methods must not, however, be misunderstood as eschewing law-based reasoning. Even where the judge is primarily striving for an appropriate rather than a legalistic judgement, his decision and his reasoning are still bound to the criteria laid down by law.[70] Neumann further asserts that the development from a "covert" to an "open" form of judicial reasoning, as demanded and

[63]Sambuc, p. 139.

[64]Deckert, 'Auslegung', p. 481; on the hierarchy question, see also Larenz, p. 345.

[65]Deckert, 'Auslegung', p. 481; see also Engisch, p. 101.

[66]See for example Rolf Wank, *Grenzen richterlicher Rechtsfortbildung*.

[67]Rüthers, *Rechtstheorie*, p. 214.

[68]Rüthers, *Rechtstheorie*, p. 214.

[69]Neumann, *Argumentationslehre*, p. 112.

[70]Neumann, *Argumentationslehre*, p. 6.

to some degree already diagnosed by Kriele[71] and Esser[72] has already taken place to an astoundingly broad extent.[73]

In Switzerland, Arthur Meier-Hayoz had previously dealt with the role of the judge in his habilitation thesis "*Der Richter als Gesetzgeber*" (The judge as legislator; 1951). In this he drew attention to the value problem, pointing out particularly that in the concrete case it was difficult to draw a line between statutory and judicial value-assessment:

> Considered in the abstract it can [...] be said that the demarcation criterion between interpretation of law and supplementation of law resides in the fact that the decisive means-end aspects can be derived from the law in the former case and in the latter case must be found by the judge in his own appraisal of the interests. In the concrete case, however, where does this borderline lie between the goal defined by statute and the goal to be defined by the judge? *The distinction of principle between statutory and judicial value-assessment often vanishes when applied to concrete cases, and much of the time can barely be drawn.*[74]

Further, in his commentary on Art. 1 of the Swiss Civil Code, in addition to the grammatical, systematic, teleological and historical element Meier-Hayoz also postulates consideration of the "*realistic element*" in interpretation.[75] Although the field of vision seems somewhat narrowly confined to the question of practicability,[76] the realistic argument represents an aspect of impact analysis:

> Consideration must be given to the realia, to the actual circumstances in which the statute is rooted and which it is intended to regulate, those of the material-physical and those of the intellectual-spiritual world: economy and science, nature and technology, customs and traditions, and above all society's views and values. [...] The interpretation must also be realistic in the sense that it [...] strives for easy realizability (practicability) of the law.[77]

Furthermore, mention should be made of the consideration of equity ("*Billigkeit*"), referred to explicitly in Article 4 of the Swiss Civil Code. The equity decision is intended to correct any outrageous outcome of the judgement in the

[71] See Martin Kriele, 'Offene und verdeckte Urteilsgründe'. According to Kriele, p. 117, for the legitimacy of the judicial decision the "prerequisite is that legal foundational research successfully exposes the covert judgements underlying the discussion, i.e. the basis for assessing reasonableness." (Own trans.).

[72] See Josef Esser, *Juristisches Argumentieren im Wandel des Rechtsfindungskonzepts unseres Jahrhunderts*. Esser complains that the grounds for judgements hardly reflected the course of the argumentation (p. 9), and calls for the use of "rhetorical means of open discourse and pre-dogmatic arguments" (p. 31).

[73] Neumann, *Argumentationslehre*, p. 117.

[74] Meier-Hayoz, *Richter*, p. 58 (own trans., author's emphasis).

[75] Meier-Hayoz, *BK*, Art. 1 N 210 ff. Id., *BK* Art. 1 N 179, relying on Friedrich Carl von Savigny, who distinguished between the four interpretative elements of "grammatical", "logical", "historical" and "systematic", and adding to them the "teleological element" and the demand to consider the "internal value of the outcome". See von Savigny, pp. 213 ff., pp. 216 ff. and p. 225.

[76] On this, cf. Mathis and Anderhub, pp. 306 ff.

[77] Meier-Hayoz, *BK*, Art. 1 N 211 and N 213 (own trans.).

individual case.[78] Here again, the impact of the judgement – if only for the parties concerned – is the operative criterion.

Furthermore, Rhinow has stood up for consequentialism from an early stage. Decisions of the legislature and judiciary were, in his view, not only legitimized through procedures but also through their alignment with and continuous review against criteria of correctness:

> The structural openness of the system of norms and the corresponding integration of justice considerations and elements of social reality within the process of legal realization do not relieve the legal practitioner of responsibility for consequences but, on the contrary, make the consideration of particular norm-relevant consequences of his decision an actual duty.[79]

He concludes that impact analysis is not only an instrument for the rationalization of rule-making, but is developing into a central legitimation factor which judge-made law, in the absence of other means of legitimation – such as democratic legitimation from the legislature – depends upon. Thus, consequentialism has a function as a postulate for overcoming a structural deficit in the legitimation of judge-made law.[80]

Biaggini notes that the Swiss doctrine of method as well as Swiss practice affirms consideration of the general impact of judgements. The evaluation of results, i.e. the concern to ensure a reasonable outcome, rightly belongs to the recognized rules of interpretation. The precedent effect of every judicial decision also suggests not only acquiescing to consequentialism but making it the judge's duty.[81] However, Biaggini is another who refers to the problem of overburdening the courts with the determination of general facts and the possible consequences of law-developing rulings, since the procedural rules of the courts are not set up to mobilize extralegal expertise.[82] The questionable issue is not therefore whether the judge should be allowed to make reference to consequences, *but whether he should be allowed to develop the law if he is unable to assess the consequences*.[83]

More recently in Switzerland, Feller in particular has engaged with impact assessment in judicial practice.[84] Other than in the judicial development of law, he also locates certain consequentialist approaches in the different interpretation elements, particularly the teleological element.[85] In addition, he comes to the

---

[78] On this, cf. Mathis and Anderhub, pp. 302 ff.

[79] Rhinow, p. 256 (own trans.).

[80] Rhinow, pp. 256 f.

[81] Biaggini, pp. 395 f.

[82] Biaggini, p. 395.

[83] Biaggini, p. 396.

[84] For a concise synopsis, see also Schluep, N 2954 ff.

[85] Feller, pp. 11 ff. and p. 131. Others, e.g. Koch and Rüßmann, pp. 230 ff., want to locate consequentialism within objective-teleological interpretation. Deckert, *Folgenorientierung*, p. 55, on the other hand sees no place for consequentialist interpretation in the classic canon of methods, but notes that in practice, at least in the guise of the *ratio legis*, consequence analyses are indeed undertaken.

significant conclusion that if the statute permits several solutions, the judiciary should consider the impacts of its decision in analogous application of Art. 1 para. 2 of the Swiss Civil Code.[86] This states that impact considerations should be taken into the body of argumentation particularly when a choice has to be made between different, contradictory interpretations.[87] As in Germany, prevailing doctrine and adjudication reject a firm hierarchy of interpretation elements. The Swiss Federal Court calls this "*pragmatic methodological pluralism*":

> Interpretation of the law has to be guided by the thought that wording alone does not constitute the legal norm, but only the law itself, understood and concretized with reference to facts. What is required is the factually correct decision within the normative framework, aimed at a satisfactory outcome derived from the *ratio legis*. In this regard, the Federal Court adheres to a pragmatic methodological pluralism and deliberately refrains from subjecting the individual interpretation elements to a hierarchical order of priorities.[88]

The consequences argument could thus serve as a *meta rule* for determining the hierarchy of the nominally equal-ranking, interpretation elements in the adjudicated case.

The Swiss constitutional law textbook "*Schweizerisches Bundesstaatsrecht*" might accurately sum up the body of opinion in Switzerland:

> The weighting of the different interpretation elements in the individual case contains an *element of valuation*. In this regard, the adjudicating body must also have regard for the outcome of the interpretation: in choosing from among the available interpretations, one of its aims must be a *satisfactory, reasonable and practicable outcome*. This is the remit and the endeavour – whether consciously or unconsciously, overtly or covertly – of every responsible legal adjudicator. If the adjudicating authorities want to be more than mere subsumption machines, they bear a share of the responsibility for meaningful judgements.[89]

The "outcome of the interpretation"[90] certainly might be a reference to the legal consequences and the directly linked micro-level real consequences, but

---

[86]Feller, p. 134.

[87]Gächter, p. 188.

[88]BGE 134 IV 297, E. 4.3.1, p. 302 (own trans.).

[89]Häfelin, Haller and Keller, N 135 (own trans.).

[90]The reverse deduction from the result is not unproblematic. Clearly there is no skill in hitting the bull's-eye by painting the rings around the bullet hole. It is possible, however, that the "great skill" lies precisely in using the application and construction of dogmatic devices to steer the result in a certain direction with apparently logical stringency. Cf. Kriele, p. 110, with the demand for disclosure of the true grounds for decisions (pp. 116 f.). Coles, p. 185, even claims that regardless of whether the judge refers directly to consequences or relies on dogmatic concepts, institutions or theories which, in their current meaning, are determined by analysis of consequences, the influence of impact considerations on judicial decisions can be proven. Seiler, *Einführung*, p. 222, considers it to be a perfectly legitimate procedure, initially to decide a question intuitively and subsequently to examine the result on the basis of the interpretation elements after the manner of a rationality and plausibility test. The important thing is openness to departing from the intuitively-found result rather than insistence on upholding it at all costs. And Sambuc, p. 111, rightly points out that analyses of consequences should not open up a pathway to a desirable result which circumvents established statutory authorities. On the danger of disregarding this principle, see Bernd Rüthers, *Die unbegrenzte Auslegung. Zum Wandel der Privatrechtsordnung im Nationalsozialismus*.

the macro-level real consequences are not necessarily ruled out either.[91] Thus, Forstmoser and Vogt demand that consideration of impacts be undertaken as part of interpretation, also giving regard to the precedent effect of the judgement and its wider implications beyond the individual case.[92] Wherever a court finds itself operating within the scope of permissible judicial law-finding, it should embark on consideration of the consequences, also giving due regard to the breadth of impact of an envisaged judgement in its deliberations.[93] Kramer calls for the judge, if his decision is not or only very vaguely "conditionally programmed" by the legislator, not to hide behind imaginary "fundamental principles" (*Grundwertungen*) of the legal order or fictional elements of legal provisions (*Tatbestandsfiktionen*) but should, just like the legislator, consciously justify these in autonomous terms. This is held up not only as a requirement of *methodological honesty* but also as the only way that a judicial ruling can become *accessible to discussion and acceptance.*[94]

As already mentioned, there is a long tradition of consequentialism in the Anglo-American world. An early proponent was Oliver Wendell Holmes in his essay "The Path of the Law" (1897), in which he strongly advocated consideration of consequences in the application of law, and turned his back on rigid formalism:

> Behind the logical form lies a judgment as to the relative worth and importance of competing legislative grounds, often an inarticulate and unconscious judgment, it is true, and yet the very root and nerve of the whole proceeding. You can give any conclusion a logical form.[95]

Holmes wanted to overcome rigid formalism and traditional dogmas, and in their place, urged consideration of the goals and the choice of means to attain these goals. Holmes was also thinking particularly of economic analysis methods:

> I look forward to a time when the part played by history in the explanation of dogma shall be very small, and instead of ingenious research we shall spend our energy on a study of the ends sought to be attained and the reasons for desiring them. As a step toward that ideal it seems to me that every lawyer ought to seek an understanding of economics.[96]

Hart even talks about a "revolt against formalism" in which Holmes was joined by the philosopher John Dewey and the economist Thorsten Veblen.[97] Hart reports that lawyers in America consequently waved goodbye to the idea that legal thought

[91]On the concepts of micro-level and macro-level real consequences, see the beginning of Section 3.

[92]Forstmoser and Vogt, § 19 N 113.

[93]Forstmoser and Vogt, § 19 N 115.

[94]Kramer, p. 218.

[95]Holmes, p. 466.

[96]Holmes, p. 474.

[97]Hart, p. 130. In the continental European system, however – despite the Free Law School *(Freirechtsschule)* and Jurisprudence of Interests *(Interessenjurisprudenz)* – these ideas have not prevailed.

was independent of policy and social reality.[98] "The Path of the Law" was published one year before William James' ground-breaking first essay on pragmatism, "Philosophical Conceptions and Practical Results" (1898), according to which the only test of an idea was to examine its practical consequences.[99] This essay marked the beginning of the pragmatic movement in the United States, the term "pragmatism" being attributed to Charles Sanders Peirce.[100]

John Dewey defended the view that consequentialism increased legal certainty rather than diminishing it. In view of the rapid pace of social change, he believed, it is difficult to fit new facts into old categories. The consequence of this was irrationality and unpredictability in legal decisions:[101]

> [T]o claim that old forms are ready at hand that cover every case and that may be applied by formal syllogizing is to pretend to a certainty and regularity which cannot exist in fact. The effect of the pretension is to increase practical uncertainty and social instability.[102]

Therefore, in order to address this problem, legal decisions should be made with their consequences in mind. This pragmatic instrumentalism gained ground in the USA and culminated in Karl Llewellyn's "The Common Law Tradition" (1960).[103]

More recently, Neil MacCormick is another who refers to the necessity of considering consequences. He rejects any one-sided mode of deliberation, whether it be informed only by consequences or only by the "correctness" of the decision. Instead he favours a middle way, in which consequences have a certain role to play.[104]

> So I reject both extremes and entertain only the middle view that some kinds and some ranges of consequences must be relevant to the justification of decisions. [...] I conclude that some element of consequentialist reasoning must be present in any sound decisionmaking process, in any satisfactory mode of practical deliberation.[105]

Although a judicial judgement is based on legal principles, this alone is not sufficient for complete legitimation of the judgement. It also has to be tested against various other factors including its consequences.[106] For MacCormick, the consequentialist argument seems to relate more to normative consequences, i.e. the repercussions of a judgement on later judgements and on other legal rules.[107] He mentions a number of criteria that should be applied:

---

[98] Horwitz, p. 142.

[99] James, p. 434; Horwitz, p. 142.

[100] James, p. 406.

[101] MacCormick, 'Legal Decisions', pp. 241 f.

[102] Dewey, p. 26.

[103] MacCormick, 'Legal Decisions', p. 242.

[104] MacCormick, 'Legal Decisions', pp. 239 f.

[105] MacCormick, 'Legal Decisions', p. 240.

[106] MacCormick, *Legal Reasoning*, p. 250.

[107] Rudden, pp. 193 f.

> It involves multiple criteria, which must include at least 'justice', 'common sense', 'public policy', and 'legal expediency'.[108]

But on that basis, particularly in the stipulation of public policy and expediency as criteria, real consequences are brought within the scope of the test as well.

3.3 Implications for Legal Practice

As this analysis shows, a blanket rejection of the consideration of consequences in legal adjudication is unconvincing. Interestingly, the discussion is played out in the different legal cultures of the German-speaking and the Anglo-American world in a similar way. Application of the law is not merely application of pre-existing rules but frequently also has a law-creating component. *It is therefore impossible to maintain a strict separation between legal and legal policy arguments.*

In summary, it can therefore be concluded that whenever appliers of the law find themselves making law (*modo legislatoris*), they are subject to the same *responsibility for consequences* as the legislator, a responsibility that can only be fulfilled by carrying out the relevant impact assessments. This applies specifically to judicial development of law and – at least in hard cases – to the interpretation of legal norms as well, particularly when a choice has to be made between two contradictory interpretations.

If, as a result, consequential arguments flow into the judicial decision process, it follows that the corresponding *grounds for decisions must be disclosed.* This is not only a requirement of methodological honesty and transparency, but is also necessary because it makes judgements easier to understand and discuss. In this way, impact analysis may not only improve the quality of judgements but also *increase the legitimacy of the application of the law.*

4 The Example of the "Hand Rule"

One famous example of the use of economic reasoning tools in the application of law is known as the "Hand rule", named after Learned Hand, an American federal judge. In a judgement delivered in 1947, he formulated an economic method for determining the efficient standard of care, which has passed into the literature as the *Learned Hand formula* or the *Hand rule*. The case on which he was ruling hinged on whether the owner of a barge could be made liable for leaving it unattended for several hours. During this time the barge had broken free from its mooring and gone on to collide with another vessel. Judge Hand explained in his judgement:

> [T]here is no general rule to determine when the absence of a barge or other attendant will make the owner of the barge liable for injuries to other vessels if she breaks away from

---

[108] MacCormick, *Legal Reasoning*, pp. 252 f.

> her moorings. [...] It becomes apparent why there can be no such general rule, when we consider the grounds for such a liability. Since there are occasions when every vessel will break from her moorings, and since, if she does, she becomes a menace to those about her, the owner's duty, as in other similar situations, to provide against resulting injuries is a function of three variables: (1) The probability that she will break away; (2) the gravity of the resulting injury, if she does; (3) the burden of adequate precautions.[109]

In stating the reasons for the judgement, Judge Hand concretized these points in a mathematical formula. If B represents the costs of the injurer's precautions, P the probability of damage and L the likely magnitude of damage, then there is a tortious liability for negligence as long as $B < P * L$. Under the Hand rule, liability for negligence begins at precisely the point where the expected value of damage exceeds the cost of avoidance. If the cost of avoidance would amount to more than the expected value of the potential damage, on the other hand, then no such liability arises for failing to take the corresponding precautions. In more general terms, according to the Hand rule, a particular action is only required if it is efficient, i.e. if it generates more benefits than costs for society.[110]

4.1 The Consequences Paradox

At first glance, the Hand rule appears to be an extremely practical rule for determining negligence, because the court can set the standard of care individually for each liability case, so that due regard is paid to the special circumstances of the particular case.[111] This practice allows the potential injurer to better gauge the likelihood of liability for negligence, and to make the appropriate choice between desisting from some dangerous activity or, instead, spending a proportionate sum on precautions. In this sense, the Hand rule is an efficient liability rule because overall it contributes to minimizing the societal costs of damages and the avoidance of damages.

Problems can arise, however, when it comes to the question of how to quantify the relevant consequences. The court has to assess *ex post* whether the injurer has assessed the risk of his behaviour *ex ante* and has spent a proportionate sum on precautions. It can be decisive for liability whether only the direct consequences of the particular case (i.e. the micro-level real consequences) are taken into consideration, or whether wider-ranging consequences for society (i.e. the macro-level real consequences) are factored in. This can lead to a paradox, as the following example illustrates.[112]

Imagine that a car driver ignores a red traffic signal in order to get a person with a life-threatening injury to the hospital in time. The costs of the accident thus caused are 40. The benefit arising from saving the injured person amounts to 50.

---

[109]United States v. Carroll Towing Co., 159 F.2d 169, 173 (2d Cir. 1947).

[110]For a more detailed discussion, see Mathis, pp. 97 ff.

[111]Cooter and Ulen, pp. 351 f.

[112]The following example presupposes fault-based liability on the part of the driver.

On the basis of these figures, crossing the red light and saving the injured person is efficient from the driver's point of view. Applying the Hand rule to these data, the court would not deem the driver's behaviour to be negligent, and would find no liability.[113]

However, if we now assume that this court judgement brings forth further-reaching consequences for society as a whole – for example, if it leads to a general deterioration in compliance with traffic laws by other drivers – then in fact the court ought to take these consequential costs into account, both in its calculation and in reaching its decision. If these consequential costs to society amount to 20, for instance, the total costs would be 60 and would then outweigh the benefit of 50. Under the Hand rule, the driver of the vehicle would then have breached his duty of care, and would be held liable.[114]

Herein lies the *paradox of consequences*: as shown, under a *judgement A*, which only takes account of the costs arising from accident, the injurer would not be liable. At the same time, a consequence of the liability-dismissing judgement A would be a general drop in compliance with the rules of the road, for which reason society would incur consequential costs of 20 in total. Now the court might want to take these societal costs into consideration as well, and would then determine that the new total costs of 60 were now higher than the anticipated benefit of 50. On that basis, the court applying the Hand rule ought to come to the conclusion that the injurer had acted negligently and that it should pronounce *judgement B*, holding him liable. If the court went on to review this judgement B with regard to its consequential costs to society, it would now find that there were none. The liability-finding judgement B – unlike judgement A – would not result in worse compliance with traffic rules. In the absence of consequential costs to society, the judgement would revert back to the liability-dismissing judgement A, whereupon the ensuing consequential costs to society from this judgement A would justify reinstatement of the liability-finding judgement B, and so on. This instance of circular logic stems from the fact that the consequential costs are dependent on the given judgement, while the given judgement in each case depends on whether or not society incurs consequential costs.[115]

To solve this problem, Fletcher proposes a strict differentiation between the "consequences of the act being judged" and the "consequences of the act of judging".[116] Based on this distinction, he postulates two solutions: the societal consequences of the individual case (i.e. the macro-level real consequences) should not be considered when determining the standard of due care, unless they are extraordinarily high.

---

[113]Fletcher, p. 191.

[114]Fletcher, p. 191.

[115]Note that this is a constructed example, however, and the assumption that judgement A would lead to generally lower compliance with the rules of the road is especially questionable. If drivers only cross red lights *in comparable instances*, the problem does not arise.

[116]Fletcher, p. 193.

Alternatively, the more radical solution would be never to consider the macro-level real consequences when assessing a concrete case.[117]

4.2 The Bilateralism Critique

An objection to the Hand rule is that this liability rule rests solely on the criterion of efficiency, in that it attempts *ex ante* to define the optimum extent of precautions for potential injurers and injured parties and to offload the damage onto the party which could have avoided it with the minimum burden.[118] The aim is thus to set the right incentives for cost-effective behaviour and hence to maximize social welfare.[119] In this sense, the Hand rule is intended to have a purely deterrent function (preventative effect).

Coleman levels the criticism, known as the *bilateralism critique*, that the economic analysis of law is incapable of explaining the normative relationship between the injured party and the injurer: all it is doing is applying *ex ante* analysis of hypothetical damages-cases from the viewpoint of cost and risk minimization. In point of fact, however, a court has to rule *ex post* on real damages-cases involving two very concrete parties who stand in a normative relationship with one another based on the case at issue:[120]

> The problem that confronts economic analysis, or any entirely forward-looking theory of tort law, is that it seems to ignore the point that litigants are brought together in a case because one alleges that the other has harmed her in a way she had no right to do. Litigants do not come to court in order to provide the judge with an opportunity to pursue or refine his vision of optimal risk reduction policy.[121]

For Coleman it is the concept of corrective justice that best explains the relationship between the injurer and the injured party.[122] But instead of taking its orientation from corrective justice, which is predicated on the bilateral nature of the legal relationship, economic analysis of law pursues a social goal, that of promoting efficiency.[123]

---

117 Fletcher, pp. 193 f.

118 Coleman, *Practice*, p. 14. Calabresi, pp. 136 ff., refers in this context to the "cheapest cost avoider".

119 Coleman, *Practice*, p. 13. For a more detailed discussion, see Mathis, pp. 166 ff.

120 Coleman, *Practice*, pp. 16 ff.

121 Coleman, *Practice*, p. 17.

122 Similar reasoning is followed by Ernest J. Weinrib, *The Idea of Private Law*; Benjamin Zipursky, 'Rights, Wrongs, and Recourse in the Law of Torts', and Martin Stone, 'On the Idea of Private Law'. On the same theme, see Jules Coleman, 'Tort Law and the Demands of Corrective Justice', and Stephen R. Perry, 'Comment on Coleman: Corrective Justice'.

123 Associated with this, according to Coleman, is the unassailable belief in the state as the engine of social change. Coleman, 'Costs', p. 344.

Interestingly, Dworkin considers the Hand rule to be compatible with his theory and disputes that it is used in pursuit of a collective goal. As he frames it, it is just a mechanism for the reconciliation of rights:

> Since Hand's test [and similar arguments] are methods of compromising competing rights, they consider only the welfare of those whose abstract rights are at stake. They do not provide room for costs or benefits to the community at large, except as these are reflected in the welfare of those whose rights are in question. [. . .] Hand's formula, and more sophisticated variations, are not arguments of that character; they do not subordinate an individual right to some collective goal, but provide a mechanism for compromising competing claims of abstract right.[124]

If one goes along with Dworkin's model, this would rebut the bilateralism critique – at least for the Hand rule – because, clearly, no collective goal whatsoever would be pursued. Not only that, but at the same time the standard of due care would be defined with reference to the benefits and costs of the affected parties only (i.e. the micro-level real consequences), which would also avoid the paradox of consequences mentioned above. Unfortunately, though, Dworkin's reasoning is unpersuasive, for even if the only the benefits and costs of the affected parties were taken into account, there is no rationalizing away the fact that the Hand rule is pursuing allocative efficiency – albeit imperfectly, since it does not include all social benefits and costs – as a collective goal.

4.3 Approaches in Swiss Liability Law

Swiss liability law distinguishes between fault-based liability *(Verschuldenshaftung)* and objective or causal liability *(Kausalhaftung)*.[125] The most important non-contractual fault-based liability is provided for in Article 41 of the Swiss Code of Obligations, according to which the liability rests with the party who unlawfully, whether wilfully or negligently, causes damage to another.

In Swiss doctrine and judicial practice, both intent and negligence are subsumed within the concept of fault. According to the traditional view, negligence is understood to mean that a person of sound mind has not acted with the degree of diligence that the average reasonable person would have exercised in the same circumstances. What this definition does not determine is how diligently an average reasonable person would act in these precise circumstances.[126] However, before a court can deliberate on the matter of fault in a concrete case, it needs a standard of care by which to define reasonable behaviour.

The Hand rule is one possible reference that might be used in order to determine how the average reasonable person would behave in the same situation. If so, the concept of the average reasonable person would correspond to the economic

[124] Dworkin, *TRS*, pp. 99 f.
[125] Roberto, § 3 N 34.
[126] Bieri, p. 289.

construct of *homo economicus*.[127] It is further noted that economic considerations in Swiss liability law are not entirely new. For example, despite the provision on causal liability in Article 58 of the Code of Obligations, the duty of diligence of a building or construction-owner is not unlimited, since the acceptability of maintenance measures to prevent construction defects is determined by comparing the corresponding costs with the benefit produced.[128]

Kramer argues emphatically in favour of considering economic factors when concretizing the key liability-law concept of negligence, referring explicitly to the Hand rule.[129] According to the Swiss federal judge Hansjörg Seiler and to Laurent Bieri, the Hand rule could already be applied *de lege lata* in Swiss liability law.[130] If the courts do not follow this opinion, it would be necessary to amend Article 41 of the Code of Obligations *de lege ferenda* as required, in order to anchor the Hand rule explicitly in statute as a standard of care for the assessment of fault.

Acknowledgment With thanks to Balz Hammer, MLaw, for his valued assistance.

Bibliography

Auer, Marietta, *Materialisierung, Flexibilisierung, Richterfreiheit* (Tübingen, 2005)

Biaggini, Giovanni, *Verfassung und Richterrecht. Verfassungsrechtliche Grenzen der Rechtsfortbildung im Wege der bundesgerichtlichen Rechtsprechung* (Basel and Frankfurt a.M., 1991)

Bieri, Laurent, 'La faute au sens de l'article 41 CO, Plaidoyer pour une reconnaissance explicite de la "règle de Hand"', in *Schweizerische Juristen-Zeitung (SJZ)*, Vol. 103 (2007), pp. 289 ff.

Bittner, Claudia, *Recht als interpretative Praxis. Zu Ronald Dworkins allgemeiner Theorie des Rechts* (Berlin, 1988)

Böhret, Carl and Konzendorf, Götz, *Leitfaden zur Gesetzesfolgenabschätzung, ed. Bundesministerium des Innern (German Federal Ministry of the Interior)* (Berlin, 2000)

Böhret, Carl and Konzendorf, Götz, *Handbuch der Gesetzesfolgenabschätzung (GFA)* (Baden-Baden, 2001)

Bundesministerium des Innern (German Federal Ministry of the Interior), *Der Mandelkern-Bericht – Auf dem Weg zu besseren Gesetzen* (Berlin, 2001)

Calabresi, Guido, *The Costs of Accidents. A Legal and Economic Analysis* (New Haven, 1970)

Coleman, Jules, 'Tort Law and the Demands of Corrective Justice', in *Indiana Law Journal*, Vol. 67 (1991/92), pp. 349 ff.

Coleman, Jules, *The Practice of Principle, In Defence of a Pragmatist Approach to Legal Theory* (Oxford et al. 2001; cited as: *Practice*)

Coleman, Jules, 'The Costs of The Costs of Accidents', in *Maryland Law Review*, Vol. 64 (2005), pp. 337 ff. (cited as: 'Costs')

Coles, Christina, *Folgenorientierung im richterlichen Entscheidungsprozeß. Ein interdisziplinärer Ansatz* (Frankfurt a.M. etc., 1991)

---

127 On this concept, see Mathis, pp. 21 ff.

128 Rey, N 1057, with further references to the doctrine and judicial practice. According to Roberto, § 11 N 401, an upkeep defect *(Werkmangel)* is always found when the costs of the safety measures are lower in total than the scale of potential damage multiplied by the probability of the occurrence of such damage.

129 Kramer, p. 235.

130 Seiler, 'Sicherheit', p. 150; Bieri, p. 296.

Cooter, Robert and Ulen, Thomas, *Law and Economics* (5th edn., Boston etc., 2008)
Deckert, Martina R., *Folgenorientierung in der Rechtsanwendung* (Munich, 1995; cited as: *Folgenorientierung*)
Deckert, Martina R., 'Zur Einführung – Die folgenorientierte Auslegung', in *Juristische Schulung (JuS)* 1995, pp. 480 ff. (cited as: 'Auslegung')
Deckert, Martina R., 'Praktische Durchführbarkeit folgenorientierter Rechtsanwendung – Auf dem Weg zu einer folgenorientierten Rechtswissenschaft', in Hagen Hof and Martin Schulte (eds.), *Wirkungsforschung zum Recht III: Folgen von Gerichtsentscheidungen* (Baden-Baden, 2001), pp. 177 ff.
Dewey, John, 'Logical Method and Law', in *The Cornell Law Quarterly*, Vol. 10 (1924/25), pp. 17 ff.
Dietrich, Matthias, *Folgenabschätzung von Gesetzen in Deutschland und Großbritannien. Ein Vergleich zum Stand der Institutionalisierung* (Munich, 2009)
Dworkin, Ronald, *Taking Rights Seriously* (Cambridge 1977; cited as: *TRS*)
Dworkin, Ronald, *A Matter of Principle* (Oxford, 1986; cited as: *MP*)
Dworkin, Ronald, *Law's Empire* (Cambridge, 1986; cited as: *LE*)
Eidenmüller, Horst, *Effizienz als Rechtsprinzip* (3rd edn., Tübingen, 2005)
Engisch, Karl, *Einführung in das juristische Denken*, edited and revised by Thomas Würtenberger and Dirk Otto (10th edn., Stuttgart, 2005)
Esser, Josef, *Vorverständnis und Methodenwahl in der Rechtsfindung. Rationalitätsgrundlagen richterlicher Entscheidungspraxis* (2nd edn., Frankfurt a.M., 1972; cited as: *Vorverständnis*)
Esser, Josef, 'Möglichkeiten und Grenzen des dogmatischen Denkens im modernen Zivilrecht', in *Archiv für civilistische Praxis (AcP)*, Vol. 172 (1972), pp. 97 ff. (cited as: 'Möglichkeiten')
Esser, Josef, *Juristisches Argumentieren im Wandel des Rechtsfindungskonzepts unseres Jahrhunderts* (Heidelberg, 1979)
Feller, Urs, *Folgenerwägungen und Rechtsanwendung* (Zurich, 1998)
Fletcher, George P., *Basic Concepts of Legal Thoughts* (New York and Oxford, 1996)
Forstmoser, Peter and Vogt, Hans-Ueli, *Einführung in das Recht* (4th edn., Bern, 2008)
Gächter, Thomas, 'Praktikabilität und Auslegung im Sozialversicherungsrecht', in *Schweizerische Zeitschrift für Sozialversicherung und berufliche Vorsorge (SZS)*, Vol. 53 (2009), pp. 182 ff.
Greenawalt, Kent, 'Discretion and Judicial Decision: The Elusive Quest for the Fetters that Bind Judges', in *Columbia Law Review*, Vol. 75 (1975), pp. 359 ff. (cited as: 'Discretion')
Greenawalt, Kent, 'Policy, Rights, and Judicial Decision', in *Georgia Law Review*, Vol. 11 (1977), pp. 991 ff. (cited as: 'Policy')
Grimm, Dieter, 'Entscheidungsfolgen als Rechtsgründe: Zur Argumentationspraxis des deutschen Bundesverfassungsgerichts', in Gunther Teubner (ed.), *Entscheidungsfolgen und Rechtsgründe. Folgenorientiertes Argumentieren in rechtsvergleichender Sicht* (Baden-Baden, 1995), pp. 139 ff.
Häfelin, Ulrich, Haller, Walter and Keller, Helen, *Schweizerisches Bundesstaatsrecht* (7th edn., Zurich, Basel and Geneva, 2008)
Hanisch, Andrea, *Institutionenökonomische Ansätze in der Folgenabschätzung der Europäischen Kommission* (Berlin, 2008)
Harris, James William, *Legal Philosophies* (2nd edn., Oxford, 2004)
Hart, Herbert Lionel Adolphus, 'American Jurisprudence Through English Eyes: The Nightmare and the Noble Dream', in id., *Essays in Jurisprudence and Philosophy* (Oxford, 1983), pp. 123 ff.
Hiebaum, Christian, 'Zur Unvermeidlichkeit von Zielsetzungen im juristischen Diskurs – Ein Argument gegen Dworkin', in *Archiv für Rechts- und Sozialphilosophie (ARSP)*, Vol. 88 (2002), pp. 86 ff.
Hoffmann-Riem, Wolfgang, 'Methoden einer anwendungsorientierten Verwaltungsrechtswissenschaft', in Eberhard Schmidt-Aßmann and Wolfgang Hoffmann-Riem (eds.), *Methoden der Verwaltungsrechtswissenschaft* (Baden-Baden, 2004) pp. 9 ff.

Holmes, Oliver Wendell, 'The Path of the Law', in *Harvard Law Review*, Vol. 10 (1886/87), pp. 457 ff.

Hopt, Klaus, 'Finale Regelungen, Experiment und Datenverarbeitung in Recht und Gesetzgebung', in *Juristenzeitung (JZ)*, 27th year (1972), pp. 65 ff.

Horwitz, Morton J., *The Transformation of American Law 1870-1960. The Crisis of Legal Orthodoxy* (Oxford, 1992)

James, William, 'Philosophical Conceptions and Practical Results', in id., *Collected Essays and Reviews* (Boston, 1920), pp. 406 ff.

Kelsen, Hans, *Reine Rechtslehre* (2nd edn. 1960, reprint, Vienna, 2000; cited as: *RR*)

Kelsen, Hans, 'Wer soll der Hüter der Verfassung sein?', in Robert Chr. van Ooyen (ed.), *Wer soll der Hüter der Verfassung sein? Abhandlungen zur Theorie der Verfassungsgerichtsbarkeit in der pluralistischen, parlamentarischen Demokratie* (Tübingen, 2008), pp. 58 ff. (cited as: 'Verfassung')

Koch, Hans-Joachim and Rüßmann, Helmut, *Juristische Begründungslehre. Eine Einführung in Grundprobleme der Rechtswissenschaft* (Munich, 1982)

Kramer, Ernst A., *Juristische Methodenlehre* (2nd edn., Bern, Munich and Vienna, 2005)

Kriele, Martin, 'Offene und verdeckte Urteilsgründe. Zum Verhältnis von Philosophie und Jurisprudenz heute', in *Collegium Philosophicum. Studien Joachim Ritter zum 60. Geburtstag* (Basel and Stuttgart, 1965), pp. 99 ff.

Larenz, Karl, *Methodenlehre der Rechtswissenschaft* (6th edn., Berlin etc., 1991)

Lave, Lester B., 'Benefit-Cost Analysis – Do the Benefits Exceed the Costs?', in Robert W. Hahn (ed.), *Risks, Costs, and Lives Saved – Getting Better Results from Regulation* (New York and Oxford, 1996; cited as: 'Benefit-Cost Analysis')

Llewellyn, Karl N., *The Common Law Tradition: Deciding Appeals* (Boston, 1960)

Lübbe-Wolff, Gertrude, *Rechtsfolgen und Realfolgen. Welche Rolle können Folgenerwägungen in der juristischen Regel- und Begriffsbildung spielen?* (Freiburg and Munich, 1981)

Luhmann, Niklas, 'Funktionale Methode und juristische Entscheidung', *Archiv des öffentlichen Rechts (AöR)*, Vol. 94 (1969), pp. 1 ff. (cited as: 'Methode')

Luhmann, Niklas, *Rechtssoziologie 2* (Reinbek bei Hamburg, 1972; cited as: *Rechtssoziologie*)

Luhmann, Niklas, *Rechtssystem und Rechtsdogmatik* (Stuttgart etc., 1974; cited as: *Rechtssystem*)

Luhmann, Niklas, 'Juristische Argumentation: Eine Analyse ihrer Form', in Gunther Teubner (ed.), *Entscheidungsfolgen und Rechtsgründe. Folgenorientiertes Argumentieren in rechtsvergleichender Sicht* (Baden-Baden, 1995), pp. 19 ff. (cited as: 'Argumentation')

Luhmann, Niklas, *Recht und Automation in der öffentlichen Verwaltung* (2nd edn., Berlin, 1997; cited as: *Automation*)

Luhmann, Niklas, *Ausdifferenzierung des Rechts. Beiträge zur Rechtssoziologie und Rechtstheorie* (Frankfurt a.M., 1999; cited as: *Ausdifferenzierung*)

MacCormick, Neil, *Legal Reasoning and Legal Theory* (Oxford, 1978; cited as: *Legal Reasoning*)

MacCormick, Neil, 'On Legal Decisions and Their Consequences: From Dewey to Dworkin', in *New York University Law Review*, Vol. 58 (1983), pp. 239 ff. (cited as: 'Legal Decisions')

Mader, Luzius, 'Zum aktuellen Stand der Gesetzesfolgenabschätzung in der Schweiz', in Ulrich Karpen and Hagen Hof (eds.), *Wirkungsforschung zum Recht IV: Möglichkeiten einer Institutionalisierung der Wirkungskontrolle von Gesetzen* (Baden-Baden, 2003), pp. 96 ff.

Mathis, Klaus, *Effizienz statt Gerechtigkeit? Auf der Suche nach den philosophischen Grundlagen der Ökonomischen Analyse des Rechts*, 3rd edn., Berlin, 2009 (English edition: Efficiency Instead of Justice? Searching for the Philosophical Foundations of the Economic Analysis of Law, New York 2009)

Mathis, Klaus and Anderhub, Alain, 'Die Kosten der Einzelfallgerechtigkeit. Praktikabilität des Rechts aus rechtlicher und ökonomischer Sicht', in *Schweizerische Zeitschrift für Sozialversicherung und berufliche Vorsorge (SZS)*, Vol. 53 (2009), pp. 301 ff.

Meier-Hayoz, Arthur, *Der Richter als Gesetzgeber. Eine Besinnung auf die von den Gerichten befolgten Verfahrensgrundsätze im Bereiche der freien richterlichen Rechtsfindung gemäss Art. 1 Abs. 2 des schweizerischen Zivilgesetzbuches* (Zurich, 1951; cited as: *Richter*)

Meier-Hayoz, Arthur, 'Art. 1 ZGB', in id. (ed.), *Berner Kommentar. Kommentar zum schweizerischen Privatrecht, Schweizerisches Zivilgesetzbuch, Vol. 1: Einleitung und Personenrecht, 1. Abteilung: Einleitung, Artikel 1-10 ZGB* (Bern, 1966; cited as: *BK*)

Montesquieu, Charles-Louis de Secondat, *De l'esprit des lois, 2 vols.*, 1748, edited by Gonzague Truc (Paris, 1949)

Neumann, Ulfrid, *Juristische Argumentationslehre* (Darmstadt, 1986; cited as: *Argumentationslehre*)

Neumann, Ulfrid, 'Theorie der juristischen Argumentation', in Winfried Brugger and Stephan Kirste (eds.), *Rechtsphilosophie im 21. Jahrhundert* (Frankfurt a.M., 2008), pp. 233 ff. (cited as: 'Theorie')

OECD, *Regulatory Impact Analysis: Best Practices in OECD Countries* (Paris, 1997)

Ogorek, Regina, *Richterkönig oder Subsumtionsautomat? Zur Justiztheorie im 19. Jahrhundert* (Frankfurt a.M., 1986)

Ott, Walter, *Der Rechtspositivismus. Kritische Würdigung auf der Grundlage eines juristischen Pragmatismus* (2nd edn., Berlin, 1992)

Perry, Stephen R., 'Comment on Coleman: Corrective Justice', in *Indiana Law Journal*, Vol. 67 (1991/92), pp. 381 ff.

Podlech, Adalbert, 'Wertungen und Werte im Recht', in *Archiv des öffentlichen Rechts (AöR)*, Vol. 95 (1970), pp. 185 ff.

Rey, Heinz, *Ausservertragliches Haftpflichtrecht* (4th edn., Zurich, Basel and Geneva, 2008)

Rhinow, René A., *Rechtsetzung und Methodik. Rechtstheoretische Untersuchungen zum gegenseitigen Verhältnis von Rechtsetzung und Rechtsanwendung* (Basel and Stuttgart, 1979)

Roberto, Vito, *Schweizerisches Haftpflichtrecht* (Zurich, 2002)

Rottleuthner, Hubert, 'Zur Methode einer folgenorientierten Rechtsanwendung', in Frank Rotter, Ota Weinberger and Franz Wieacker (eds.), *Wissenschaften und Philosophie als Basis der Jurisprudenz*, ARSP Beiheft 13 (Wiesbaden, 1980), pp. 97 ff.

Rudden, Bernard, 'Consequences', in *The Juridical Review: The Journal of Scottish Universities*, Vol. 24 (1979), pp. 193 ff.

Rüthers, Bernd, *Die unbegrenzte Auslegung. Zum Wandel der Privatrechtsordnung im Nationalsozialismus* (Tübingen, 2005)

Rüthers, Bernd, *Rechtstheorie* (4th edn., Munich, 2008; cited as: *Rechtstheorie*)

Sambuc, Thomas, *Folgenerwägungen im Richterrecht. Die Berücksichtigung von Entscheidungsfolgen bei der Rechtsgewinnung, erörtert am Beispiel des § 1 UWG* (Berlin, 1977)

Schluep, Walter R., *Einladung zur Rechtstheorie* (Bern and Baden-Baden, 2006)

Seiler, Hansjörg, 'Wie viel Sicherheit wollen wir? Sicherheitsmassnahmen zwischen Kostenwirksamkeit und Recht', in *Zeitschrift des Bernischen Juristenvereins (ZBJV)* 2007, pp. 143 ff. (cited as: 'Sicherheit')

Seiler, Hansjörg, *Einführung in das Rechts* (3rd edn., Zurich, Basel and Geneva, 2009; cited as: *Einführung*)

Seiler, Hansjörg, *Praktische Rechtsanwendung. Was leistet die juristische Methodenlehre?* (Bern, 2009; cited as: *Rechtsanwendung*)

Sommermann, Karl-Peter, 'Folgenforschung und Recht', in id. (ed.), *Folgen von Folgenforschung. Forschungssymposium anlässlich der Emeritierung von Universitätsprofessor Dr. Carl Böhret am 16./17. November 2001* (Speyer, 2002), pp. 39 ff.

Stone, Martin, 'On the Idea of Private Law', *Canadian Journal of Law and Jurisprudence*, Vol. 9 (1996), pp. 235 ff.

van Aaken, Anne, *"Rational Choice" in der Rechtswissenschaft. Zum Stellenwert der ökonomischen Theorie im Recht* (Baden-Baden, 2001)

von Savigny, Friedrich Carl, *System des heutigen römischen Rechts*, Vol. 1 (Berlin, 1840)

Wälde, Thomas W., *Juristische Folgenorientierung. "Policy Analysis" und Sozialkybernetik: Methodische und organisatorische Überlegungen zur Bewältigung der Folgenorientierung im Rechtssystem* (Königstein, 1979)

Wank, Rolf, *Grenzen richterlicher Rechtsfortbildung* (Berlin, 1978)

Watkins-Bienz, Renée M., *Die Hart-Dworkin Debatte. Ein Beitrag zu den internationalen Kontroversen der Gegenwart* (Berlin, 2004)

Weigel, Wolfgang, *Rechtsökonomik. Eine methodologische Einführung für Einsteiger und Neugierige* (Munich, 2003)

Weinrib, Ernest J., *The Idea of Private Law* (Cambridge, 1995)

Wolf, Jean-Claude, 'Gesetzesregeln und Gesetzesprinzipien', in Steffen Wesche and Véronique Zanetti (eds.), *Dworkin in der Diskussion* (Paderborn, 1999)

Zipursky, Benjamin, 'Rights, Wrongs, and Recourse in the Law of Torts', in *Vanderbilt Law Review*, Vol. 51 (1998), pp. 1 ff.

Consequence-Based Arguments in Legal Reasoning: A Jurisprudential Preface to *Law and Economics*

Péter Cserne

One of the persistent problems surrounding the discipline of *law and economics* is the role of economic arguments in legal reasoning. The problem has been extensively discussed in the literature but has not been ultimately solved.[1] The present paper is a contribution to this ongoing discussion. The argument goes as follows. First, I will argue that insights from *law and economics*, to the extent that they claim to be directly relevant for legal reasoning, should carry a jurisprudential preface that states that this very relevance is limited and conditional. Secondly, I will introduce the concept of consequence-based reasoning and show that the typical normative claims of *law and economics* based on economic efficiency can be interpreted as consequence-based arguments of a special kind. Finally, in the analytical core of the paper, the conceivability, feasibility and desirability of the judicial appreciation of general social consequences of legal decisions will be considered. Referring to the philosophical, jurisprudential and institutional dimensions of the issue I will argue that in a modern constitutional democracy the scope of consequence-based judicial reasoning is limited mainly by the expertise of courts. A more general

This paper was written as part of the research project on *Convergence and Divergence of National Legal Systems* (project number 100-16-503) at Tilburg Law and Economics Centre, sponsored by The Hague Institute for Internationalization of Law. I am grateful to Mátyás Bódig, Alon Harel, András Jakab, Pierre Larouche, Giuseppe Martinico, Bert van Roermund, Stefan Vogenauer, and participants at the First Central and Eastern European Forum of Young Legal, Social and Political Philosophers (Silesian University Katowice), Seminar on Legal and Economic Reasoning (Université de Paris Ouest Nanterre la Défense), Post-doctoral Conference in Law and Economics (Universität Hamburg), TILEC research seminar (Tilburg), the Hamburg Lectures in Law and Economics, and the Tilburg Philosophy of Law Colloquium for helpful comments and suggestions on previous versions of the paper.

[1] See, e.g. Wald, 'Limits on the Use of Economic Analysis in Judicial Decision-making'; Ulen, 'Courts, Legislatures, and the General Theory of Second Best in Law and Economics', and in more specific contexts, Breyer, *Economic Reasoning and Judicial Review* (American constitutional law); Sibony, *Le juge et le raisonnement économique en droit de la concurrence* (French and European competition law).

P. Cserne (✉)
Tilburg Law School, Tilburg University, NL-5000 LE Tilburg, The Netherlands
e-mail: peter.cserne@uvt.nl

K. Mathis (ed.), *Efficiency, Sustainability, and Justice to Future Generations*, Law and Philosophy Library 98, DOI 10.1007/978-94-007-1869-2_2,

implication of this analysis is that the impact of *law and economics* scholarship on law can only be understood through a close look at legal reasoning in general and consequence-based arguments in particular.

1 The Jurisprudential Preface

At the outset, one may legitimately ask: why is the discussion about legal reasoning relevant at all for *law and economics* scholarship? To my knowledge, there have been very few occasions where the discipline has been directly confronted with this question. One such moment of methodological self-consciousness and philosophical self-reflection occurred during a 1993 debate between American law professors Dennis Patterson and Richard Craswell.[2]

A useful starting point for our discussion is a distinction drawn by Patterson between propositions *of* law and propositions *about* law.[3] He argues that the distinction refers to the sort of claims that can be legitimately put forward in the two cases. The first type of proposition should be justified by arguments within law, the second type instead by arguments from other disciplines, such as philosophy, social psychology, theology or economics. In order to explain this distinction, Patterson draws an analogy between the practice of law and the practice of riding a bicycle.[4] A proposition *of* bicycling is that "Moving the pedals in a clockwise fashion causes the cycle to move forward." A proposition *about* bicycling is that, under normal circumstances, it is beneficial for the cardiovascular system. Patterson's point is that arguments supporting the first kind of propositions have no relevance for the second type of propositions. Vice versa, arguments from various disciplines and discourses outside the practice, such as medical propositions about biking or economic arguments about law, have no bearing on propositions internal to the practice of bicycling or of law.

In other words, legal discourse is fully or at least substantially autonomous: it sets its own standards for what are the relevant or legitimate arguments *in* law. As Patterson concludes, if this distinction is not kept in mind and legal scholars

[2] Patterson, 'The Pseudo-Debate over Default Rules in Contract Law'; Craswell, 'Default Rules, Efficiency, and Prudence'. It should be noted that my argument is different from and not relying on the discussion about the reception of *law and economics* in legal academia, legal education and legal practice in various countries. On this, see e.g. Ackerman, 'Law, Economics, and the Problem of Legal Culture'; Symposium, 'Economic Analysis in Civil Law Countries: Past, Present, Future'; Posner, 'The Future of the Law and Economics Movement in Europe'; Dau-Schmidt and Brun, 'Lost In Translation: The Economic Analysis of Law in the United States and Europe'; Gazal-Ayal, 'Economic Analysis of Law in North America, Europe and Israel'; Garoupa and Ulen, 'The Market for Legal Innovation: Law and Economics in Europe and the United States'; Grechenig and Geltner, 'The Transatlantic Divergence in Legal Thought: American Law and Economics vs. German Doctrinalism'.

[3] Patterson, pp. 239 f.

[4] Patterson, pp. 267 f.

"attempt to justify propositions of law with the justificatory tools of other disciplines", then academic debates become "pseudo debates" where arguments are used in an illegitimate way.[5]

At first sight, Patterson's thesis about the autonomy of law presents a challenge for the inclusion of economic arguments in the practice of legal reasoning. This challenge can be answered in a number of ways. The first is surrender. It seems relatively uncontroversial that economic, as well as other external (political, philosophical, etc.) arguments can be legitimately put forward as propositions *about* law. They can be both intellectually interesting and practically useful as arguments in a discourse which is separate and distinct from legal practice. This is a discourse among outsiders: experts, critics, and reformers. Eventually, their views can reach the law indirectly, for instance by influencing the preferences or beliefs of lawmakers or legal officials.[6]

There is also a second, more ambitious line of response to Patterson. Conceding that it depends on the particular legal practice whether certain kinds of external arguments are legitimate or not, one could draw a more nuanced picture of the canon of acceptable or legitimate arguments and then see whether this canon would accommodate economic arguments in a particular type of legal practice.[7] In his own response, Richard Craswell argues that one should avoid "the jurisprudential naïveté about the ultimate connection, if any, between the [. . .] technical economic analysis and the sorts of argument that might be acceptable to courts." "[I]t is appropriate to regard each economic analysis as being limited by [a] preface" which makes clear that the efficiency effects identified in the analysis should be relevant for courts in deciding cases "*to the extent prudential arguments are relevant*" (emphasis in original).[8] If economic arguments can be recast as prudential arguments then the relevance of economic analysis for legal reasoning will depend on the acceptability of those kinds of arguments.

While the Patterson-Craswell debate drew attention to this dependence, it did not directly address the latter question: the role of prudential arguments in legal reasoning. This is the main question discussed in this paper. For the rest of this paper, I will refer to consequence-based arguments which carry virtually the same meaning as prudential arguments. Note that by focusing on consequence-based arguments, we can only determine the limits or boundaries of the domain within

[5]Patterson, p. 268.

[6]Arguably, in certain contexts this answer is the only convincing one. For instance, an economic analysis or a feminist interpretation of a doctrine of ancient Roman law can be enlightening but it is hard to imagine it as a part of ancient Roman legal practice. References to Pareto efficiency or "phallogocentrism" do not seem to fit into the ancient Roman canon of acceptable arguments. Cf. Harris, 'The Uses of History in Law and Economics'.

[7]On the canon of acceptable arguments see Honoré, pp. 64 ff.; Bell, 'Acceptability'; Cserne, pp. 11 ff.

[8]Craswell, p. 293. In his paper, Craswell actually formulated a preface that *law and economics* articles should carry but he also argued that if each article is tacitly understood to stand under this proviso, it would be superfluous to articulate it each time.

which efficiency-based arguments have to find their place. In this paper I will not address the question regarding the merits of efficiency-based arguments within this larger set of arguments.[9]

A third and more radical answer to Patterson's thesis would be to challenge the strict separation of intra- and extra-systemic arguments. Instead of taking the distinction at face value, one could inquire whether his claim about the autonomy of law is a conceptual or an empirical one, and in either case, whether it is correct.

There seem to be good reasons to assume that Patterson's claim is an empirical one. He repeatedly claims that the types of arguments which can be used to justify legal propositions are conventional.[10] Conventions of the legal profession are contingent features of legal practice: they differ throughout jurisdictions and change over time. If this is the case then one can contest whether the strict separation of the discourses *of* law and *about* law is a good description of the argumentative practices one can observe in current legal systems. In his critique of Patterson, Steven Burton argued that, at least in modern US law, there is no strict separation between legal reasoning and extra-legal insights:

> [D]ifferent discourses or language games are both dynamic and overlapping, with some in process of birth or decay, others in stages of merger, and yet others about to split like an amoeba. The familiar distinctions among politics, morality, economics and sociology, for example, originated largely in the nineteenth century and are now decaying in practice. Moreover, even when discourses are distinct, each can incorporate arguments from another. The conventions of the legal profession thus can incorporate arguments from other fields as proper legal arguments [...]. The distinction between propositions of and about law then collapses as a matter of convention.[11]

Still, Patterson's autonomy thesis is not completely indefensible as an empirical claim. It would be inaccurate to say that the distinction between intra and extralegal arguments has completely disappeared, even from today's US law. Nonetheless, it seems safe to state that during the last decades, at the price of some loss in formal rationality, certain insights from specific non-legal disciplines have become increasingly accepted as arguments within certain legal discourses.[12]

Finally, can we convincingly reconstruct Patterson's autonomy thesis as a conceptual one? Although this argument cannot be fully elaborated here, there seems to be a plausible case for such an interpretation. Along the lines of a Luhmannian version of systems theory, one could argue that any developed legal system is operationally closed in the sense that it does not admit extra-systemic arguments

---

[9]On this debate see the still enlightening overview by Kornhauser, 'A Guide to the Perplexed Claims of Efficiency in the Law'.

[10]Patterson, pp. 241 f., p. 280 n. 164.

[11]Burton, p. 307.

[12]The canon of acceptable arguments is not homogeneous but domain specific and changes over time, see Cserne, pp. 16 ff. For comparative analyses of legal reasoning, especially in statutory interpretation see MacCormick and Summers, *Interpreting Statutes. A Comparative Study*; Vogenauer, *Die Auslegung von Gesetzen in England und auf dem Kontinent*; Hage, 'Legal Reasoning'.

(information, communication) as part of the operations within the system.[13] In his 1974 book *Rechtsdogmatik und Rechtssystem*, Luhmann made two bold claims in this respect. First, from an internal or doctrinal perspective, real-world consequences of judicial decisions are irrelevant. Second, real-world consequences do matter for doctrinal scholarship, namely in the process of adaptation towards "socially adequate" doctrinal concepts. While legal decisions themselves take current doctrinal concepts and categories as given, these are continuously revised and corrected in the process of doctrinal discussion.[14] In an early critical review of Luhmann's views, Gunther Teubner argued that in post-industrial societies consequence-based arguments have a constitutive, i.e. not merely corrective role in law.[15] This claim is not merely a conceptual one. Rather, as a theoretical reconstruction, it has made the increasing openness of German jurisprudential discourse towards consequence-based arguments more comprehensible.[16]

In summary, one could respond to the conceptual interpretation of the autonomy thesis like this. To the extent that legal discourse is formally rational, it has some autonomy that sets limit to the direct reference to normative ideas or scientific insights *in* law or *as* law.[17] Nevertheless, the autonomy of law does not imply its isolation from arguments of other disciplines. Rather, it refers to the specific channels through which external arguments are integrated into legal practice. While a theoretical reconstruction of the autonomy thesis in the Luhmannian terms of operational closure or *autopoiesis* is still debatable, it could be integrated into a fully fledged model of legal systems which also accounts for the role of extra-systemic arguments in a more precise way. In particular, systems theory provides a useful understanding of how doctrinal scholarship *(Rechtsdogmatik)* or legal policy *(Rechtspolitik)* can function as filters or transformation mechanisms between the legal system, narrowly understood, on the one hand and the scientific or political system on the other.[18]

2 Legal Reasoning and the Consequences of Judicial Decisions

Experienced legal practitioners have in-depth knowledge about what kinds of arguments are acceptable within particular legal contexts. However, this internal view is sometimes confused, misleading, or incomplete. It is therefore one of the tasks of legal theory to clarify the nature of legal reasoning.

---

[13]See, e.g. Luhmann, *Rechtsdogmatik und Rechtssystem.*

[14]Luhmann, *Rechtsdogmatik*, cited in Teubner, 'Folgenkontrolle', p. 181.

[15]Teubner, 'Folgenkontrolle'.

[16]Teubner, *Entscheidungsfolgen*; Luhmann, 'Legal Argumentation'; Kirchner, 'Zur konsequentialistischen Interpretationsmethode'.

[17]On legal formalism see Kennedy, 'Formality'; 'Form'.

[18]Cf. Bell, 'Policy', p. 95 n. 2: "an 'autopoietic' perspective merely confirms the view that law is open to external influences in a significant manner."

In general, legal reasoning is "an activity conducted within more or less vague or clear, implicit or explicit, normative canons. We distinguish between good and bad, more sound and less sound, relevant and irrelevant, acceptable or unacceptable arguments in relation to [. . .] legal disputation."[19] Following Tony Honoré, I refer to these criteria as the *canon of acceptable arguments*.[20] This canon arises out of a social practice amongst the legal community. By way of socialization and introduction to the skills and knowledge of the profession, members of any given legal community become enabled to deal with ordinary situations in a particular technical way, according to a specialized linguistic terminology.[21] The canon limits decision-makers both psychologically and institutionally, by requiring them to fit "within the framework".[22]

Legal reasoning, i.e. the way judges and other legal officials justify their decisions can be related to the consequences of these decisions in several ways but not all of them are relevant for our discussion. As far as legal reasoning is concerned, consequences only matter if they are explicitly referred to in public justificatory arguments.

Courts, especially higher or constitutional courts often take decisions which have large-scale social consequences. This does not mean that judges are necessarily aware of these consequences or that if they are, their decisions will be motivated by what they expect to result from their decisions. Even if, as a matter of psychology, they are influenced by the expected consequences, they are not always willing or allowed to publicly refer to them as reasons for their decision.[23]

Furthermore, even if the legal reasoning does not openly refer to or account for such consequences, in public political discussion court decisions are typically evaluated, welcomed or criticized in terms of their alleged consequences. For instance, constitutional courts are sometimes criticized for extending the justiciability of economic and social rights "irresponsibly", i.e. by neglecting the financial burdens on the state budget arising from interpreting such rights as justiciable.[24] In other cases they are accused of being "irresponsive" to social and political needs and hiding behind legal formalism. Such public scrutiny sometimes amounts to pressure that judges should be made accountable for the consequences of their decisions. In this paper I am concerned with consequences of legal decisions only as far as judges or collegiate courts provide justificatory reasons for their rulings and among these

---

[19]MacCormick, *Reasoning*, p. 12.

[20]Honoré, p. 64.

[21]Bell, 'Acceptability'.

[22]MacCormick, *Reasoning*, p. 34.

[23]In analogy to the distinction in the philosophy of science (see e.g. Schiemann, 'Inductive Justification and Discovery. On Hans Reichenbach's Foundation of the Autonomy of the Philosophy of Science'), this difference is sometimes referred to with juxtaposing the 'context of discovery' and the 'context of justification', e.g. MacCormick, *Reasoning*, pp. 15 f.; Anderson, 'Context of Discovery, Context of Decision and Context of Justification in the Law'.

[24]See, e.g. Sajó, 'How the Rule of Law Killed Hungarian Welfare Reform'.

reasons they are allowed or required to make explicit reference to the consequences of their decision.

3 What Are Consequence-Based Arguments?

In this section I will briefly discuss what I mean by consequence-based arguments. Then I distinguish some important types of these arguments and show that efficiency-based arguments fit into one of these categories.

As a first approximation, we can contrast consequence-based reasoning with rule-following. Rule-based decision-making unavoidably faces the problem of over and/or under-inclusiveness.[25] Over and/or under-inclusiveness of rules vis-à-vis their background justification occurs when general rules are applied to heterogeneous cases or subjects. Some of those cases will be covered by the rule but not by its background justification (the rule is over-inclusive) or vice versa (the rule is under-inclusive). For instance, if a legal rule sets the age limit of full legal capacity at 18 years, the rule serves as an easily administrable way of granting legal capacity to individuals who are mature enough to take care of their dealings. However, the rule will also grant capacity to those who are over 18 but immature (over-inclusiveness) and deny capacity from those who are under 18 but sufficiently mature (under-inclusiveness).

Legal systems show a great variety with regards to how much divergence will result in the return from the rule to its background justifications. They also use different doctrinal techniques to handle this divergence. In cases where the divergence is perceived to be large and important, modern legal systems usually require, allow, or at least tolerate that judges do not apply the rule as written but return to the background justification in some way. This can be seen as the main *raison d'être* of at least some kinds of consequence-based reasoning within legal systems based on rule-based decision-making. In other words, if judges were allowed to use rule-based arguments only, justifications could not be based on direct and explicit reference to consequences.

Now we can formulate a definition for consequence-based reasoning.[26] If in deciding case C, the decision-maker finds that there is a relevant rule R which has more than one plausible interpretation (X, Y, Z, ...) the decision-maker is said to use a consequence-based argument if she justifies her decision for rule-interpretation X (instead of rule-interpretation Y or Z) with the argument that rule-interpretation X will bring about consequences which are normatively superior to the consequences brought about by the alternative rule-interpretations. The normative standard adopted by the decision-maker also determines which features of the possible rule-interpretations are relevant for the decision-making process.

[25]The classic reference is, of course, Schauer, *Playing by the Rules*.

[26]To be more precise, this is a definition of consequence-based reasoning *within a system of rule-based reasoning*.

Note that according to this definition, purposive or teleological interpretation is also consequence-based reasoning. Teleological interpretation can be "subjective" when the decision-maker evaluates the consequences of the alternative interpretations X and Y in light of the goals and values which she identifies as the historical "legislative intent" that motivated the lawmaker in adopting the rule. Alternatively, it can be "objective" when the decision-maker identifies the *ratio legis*, the purpose of the rule, independently from legislative intent and chooses the interpretation that better serves this purpose.[27]

A further well-known variant of consequence-based arguments is the judicial reference to "public policy." When judges find it appropriate to deviate from a legal rule for reasons of public policy, they evaluate the consequences of alternative rulings in light of normative standards which they identify either by evoking the function(s) of a specific domain of law to which the rule belongs or by referring to more general purposes of the law, such as the protection of legitimate expectations or the proper functioning of the judicial and political system. In some cases, they weight these values against the more direct purpose of the particular rule which they have to interpret. In this sense, public policy represents "a compromise between various values the law has to serve."[28]

Having characterized consequence-based reasoning in general, further conceptual clarity can be gained by distinguishing the types of consequences that are taken into account as justificatory reasons.

4 What Type of Consequences Matter?

As the above definition does not specify the normative standard on which consequences are measured and compared, there are as many logically possible versions of consequence-based reasoning as there are normative standards. For instance, an act-utilitarian decision-maker would use the standard of overall social utility. A judge whose sole consideration is gender equality would choose the interpretation which promotes this goal the most. The utilitarian decision-maker needs to look at potentially all, known and unknown, close and remote, consequences of the alternative rule-interpretations, rank these interpretations based on their overall consequences and choose the one that comes out best. To apply this standard literally in a judicial setting would make the role impossible to fulfil. Even a perfectly conscientious Herculean utilitarian judge, with unconstrained time and the best expertise would have to face limits of information and foresight, at least because of the inherent uncertainty of the future. While the utilitarian judge is involved in what could be called "aggregative reasoning" about consequences, the judge who is

[27]On the distinction between "subjective" and "objective" teleological interpretation in German legal theory see Koch and Rüßmann, pp. 220 ff. While at first sight the normative standard which has to be used in the evaluation of consequences seems well-determined in both variants, there is an extensive jurisprudential literature to show that the identification of both the legislative intent and the *ratio legis* behind a rule runs into serious theoretical and practical difficulties.

[28]Bell, 'Policy', p. 88.

solely concerned with gender equality has to perform a somewhat simpler task of "single-factor consequence-based reasoning".[29] Real-world legal systems typically constrain judges in their choice of normative standards. Correspondingly, the kinds of consequences that are relevant for their decision are also limited.[30]

Furthermore, we can distinguish between individual and systemic; and among the latter, between juridical and behavioural consequences. A judicial decision can refer either to consequences for the parties involved in an individual case or to large-scale or general consequences. For instance, a judge may realize that if she decides that the conduct of the defendant is identified as the "adequate" or "proximate" cause of the harm suffered by the plaintiff, rules of tort law require that a large amount of damages is awarded to the plaintiff. One consequence of this ruling could be that this payment puts an extremely high financial burden on the defendant. Is the judge allowed or mandated to take such consequences into account when deciding whether the causal link between the defendant's conduct and plaintiff's harm is legally established? Or is the judge allowed to rule that although law is on plaintiff's side, "equity" or "fairness" absolves the defendant from liability? In general, modern Western legal systems answer such questions in the negative. Sometimes, however, judges are explicitly authorized to take into account the impact of their ruling on the individual(s) involved in a particular case. For instance, judges often have to decide regarding the detention of a criminal suspect based on the likelihood that the suspect will escape or commit further crimes. When judges are authorized to base their decision on such consequential considerations they also have the duty to justify their decision with arguments related to the expected consequences of alternative rulings.

Judicial decisions are sometimes justified with reference to a broader set of consequences. Returning to the example of "adequate" or "proximate" cause, a judge may realize that a broader or narrower construction of causation would have an impact on civil liability as an institutional mechanism of damage compensation throughout the legal system. She may also realize that different interpretations of the doctrinal concepts of causation would change the distribution of wealth between various social groups such as injurers and victims or affect the level of precaution by potential injurers. When judges consider the impact of their decision on the rules of civil liability, on similar tort cases in the future, or on the conduct of potential injurers and victims, *and* justify their decision with reference to such considerations, they are said to use general or systemic consequence-based arguments.

Within this latter category of general or systemic consequences one can further distinguish between "juridical" and "behavioural" consequences.[31] When a case is decided with regard to its juridical consequences, it is decided with "reference to decisions which would have to be given in other cases if a particular ruling were

---

[29]Cane, pp. 41 f.

[30]As a matter of normative theory, there are good reasons for setting such constraints. This is, however, beyond the scope of the paper.

[31]Rudden, p. 194. In the German literature, Gertrude Lübbe-Wolff makes a similar distinction (Lübbe-Wolff, *Rechtsfolgen und Realfolgen. Welche Rolle können Folgenerwägungen in der juristischen Regel- und Begriffsbildung spielen?*).

given in the instant case".[32] In other words, the judge examines the logical implications of interpretation X or Y on other rules within the legal system, by inquiring "what sorts of conduct the rule would authorize or proscribe."[33]

A typical example of reasoning from juridical consequences is the so-called "where-will-it-all-end" or slippery slope argument.[34] This is often invoked by judges or lawyers who want to convince their audience about the superiority of interpretation X by drawing attention to the disastrous consequences of deciding the opposite way. The argument goes like this: if in this case we decide Y, in all similar cases we will also have to decide Y. If we chose interpretation Y, this would imply that other rules within the legal system would allow for or mandate something that is clearly undesirable in light of the normative standard applied in this case. As this logical implication of interpretation Y is unacceptable, therefore, we should decide X.[35]

A prominent example of this slippery slope argument was put forward by Chief Justice Marshall in *Marbury v. Madison* [5 U.S. (1 Cranch) 137 (1803)]. Arguing in favour of the judicial review of statutes, Marshall conjectured that not to allow judicial review would "subvert the very foundation of all written constitutions."[36] In other words, he made an explicit reference to the unacceptable systemic juridical consequences of the alternative ruling.

While judicial consequences are "law-immanent",[37] behavioural consequences refer to "what human behaviour the rule will induce or discourage" outside the legal system, in society at large.[38] When judges refer to behavioural consequences, they make a more or less educated guess about how certain groups of legal subjects would change their behaviour in response to a certain decision. When judges decide between alternative interpretations of a rule in light of such general social consequences, they have to imagine and compare hypothetical scenarios under the assumption that individuals will change their behaviour in a predictable way, in response to how the law would regulate their dealings. Clearly, any sensible use of such an argument is based on a number of assumptions regarding how the law influences human behaviour.

In the following, I will focus on legal arguments based on such behavioural or general social consequences. One reason for this is that they have been somewhat neglected in jurisprudential literature.[39] Another reason is that the typical arguments

---

[32]Cane, p. 41.

[33]MacCormick, 'Decisions', p. 239.

[34]Cane, p. 41.

[35]MacCormick, 'Decisions', p. 240.

[36]MacCormick, 'Decisions', p. 240.

[37]Luhmann, *Risk*, p. 59.

[38]MacCormick, 'Decisions', p. 239.

[39]Systemic juridical consequences have been thoroughly discussed in previous literature. Professor Neil MacCormick who wrote extensively on consequence-based reasoning, focused almost exclusively on these. See MacCormick, *Reasoning*, ch. 6; 'Decisions'; *Rhetoric*, ch. 6. Rudden argued that by focusing on juridical consequences, MacCormick "frames his case too narrowly" (Rudden, p. 193). Cane focuses on behavioural consequences.

of *law and economics* fall into this category. A judicial decision which is justified by the improvement in economic efficiency or welfare it is supposed to bring about is an argument based on behavioural consequences. In fact, efficiency-based arguments are a subcategory of consequence-based or prudential arguments. As mentioned above, in order to determine the proper role of economic arguments in legal reasoning, we have to enter the normative discussion about the proper role of consequence-based arguments.

5 (When and Why) Should Judges Use Consequence-Based Arguments?

Analytically, one can distinguish three levels in the discussion about the proper role of consequence-based arguments in judicial reasoning: (1) philosophical (or conceptual), (2) doctrinal (or institutional), and (3) political (or pragmatic). While the audience and the terminology of these discursive levels are different, in most real-world discussions these are combined. What is important to see is that those who favour, promote or defend consequence-based judicial reasoning have to argue for it at each level, by demonstrating that consequence-based judicial reasoning is (1) conceivable, (2) feasible, and (3) desirable. These questions have a logical sequence: it only makes sense to move to the next one if the previous one is answered positively, at least partially. Moving from the first to the second and third questions also implies an increase in the importance of contextual factors such as time, jurisdiction or domain of law.

When courts justify their decisions with reference to expected consequences, serious problems emerge. The gist of my argument in the following sections is that these problems are closely linked to the competence of courts. "Competence" is a multi-faced, technical, institutional and normative feature of adjudication. It has at least two aspects: legitimacy and expertise. These aspects are analytically separable: while legitimacy relates to judicial authority, accountability, and discretion, expertise raises institutional and technical problems of judicial decision-making.

6 Conceivability and Objections from the Nature of Adjudication

Current "legal scholarship [...] draws a crude distinction between two modes of reasoning within law – instrumental, forward-looking, or policy-oriented ways of thinking and backward-looking, principled, or rule-based doctrinal reasoning."[40] While these dichotomies obviously do not overlap completely, it is plausible to see consequence-based reasoning as an instance of the first mode: it

[40]Parker et al., p. 4.

is instrumental, forward-looking and policy-oriented legal reasoning.[41] Is this compatible with the nature of adjudication in a constitutional democracy or in other systems?

For legal philosophers the key question about legal reasoning is this: what constitutes a relevant argument in law? When they construct a theory of legal reasoning, legal philosophers are interested in how a practice of giving good public reasons in support of a decision works and how the practice itself can be justified. With regard to our specific topic, normative legal philosophy is interested in the following question: what is the place for consequences among the reasons judges can put forward in justifying their decisions?

For a long time the jurisprudential discussion on legal reasoning considered judicial decision-making to be faced with the following alternative: legal formalism, deductive reasoning, "Formal Style" on the one hand, pragmatism, consequentialism, "Grand Style" on the other.[42] Legal scholars in the German-Austrian *Freirechtsschule*, American legal realism, and many others have claimed that formalism only covers or hides discretion: it makes law unpredictable, uncertain and ultimately unjust. They argued that an open recognition of a more active role of judges would make law more predictable and also substantively closer to the normative standards that these writers claimed to be the proper guiding principles of adjudication.

Starting from the 1970s, it has become generally accepted that apart from distinguishing formal and substantive reasoning, within the second category one should further distinguish between "goal-reasons" and "rightness-reasons".[43] A "goal-reason" justifies a decision as a means or instrument for promoting or securing a state of affairs that is desirable. A "rightness-reason" justifies the decision by invoking a "sociomoral norm" that the conduct or relationship of the parties concerned is supposed to be subject to. While goal-reasons are forward-looking, rightness-reasons are backward-looking.

In Ronald Dworkin's early theory, the same dichotomy arises as a distinction between policies and principles and the corresponding types of argument which support two different types of social aims.[44] Arguments of policy are meant to support decisions about collective goals and goods. Arguments of principle are supporting goods which are realized individually, especially as protected by individual rights. Dworkin argued forcefully that the proper task of courts of law, both in a descriptive and a normative sense is to ascertain and vindicate rights. He put forward the claim

---

[41] In fact, similar dichotomies can be multiplied. One can juxtapose output-oriented and input-oriented decisions; *ex ante* and *ex post* perspective; substantive and formal rationality; standards and rules; utility and rights; *voluntas* and *ratio*; interdisciplinarity and operational autonomy; *Lebensnähe* and *Lebensfremdheit*. See Kennedy 'Formality', 'Form'; Luhmann, *Rechtsdogmatik*, 'Legal Argumentation'; Koch and Rüßmann, *Juristische Begründungslehre*; Teubner, *Entscheidungsfolgen*.

[42] MacCormick, 'Decisions', pp. 242 f.

[43] The terminology comes from Robert Summers, quoted in MacCormick, 'Decisions', p. 243.

[44] Dworkin, *Taking Rights Seriously*.

that courts (should) base their decisions solely on principles and not on policies. In short, for Dworkin, there seems to be no scope for consequence-based arguments in judicial reasoning.

Critics argue that Dworkin's distinction between policies and principles is ambiguous and his arguments against policy-based adjudication are not convincing.[45] In subsequent writings Dworkin himself made the distinction less marked, noting that "the argument that D ought to be the decision in this case in order to secure or promote aim A may just as likely be an argument of principle as an argument of policy".[46] Thus, in many cases where Dworkin would reconstruct the judge's argument as one of principle, others can reconstruct it as one of policy.

A related objection against consequence-based arguments comes from universalizability. Both Dworkin and MacCormick, as well as other legal theorists, hold the view that the activity of judging implies justification and to justify a decision means to universalize. MacCormick's philosophical claim is that as a matter of conceptual or analytical truth, a judicial decision can only be justified with universalizable arguments. Justificatory reasons have to be universalizable in the sense that they should apply to each and every case which is sufficiently similar in relevant aspects to the one under scrutiny. According to MacCormick, this requirement of universalizability follows from the concept of formal justice that "like cases should be treated alike".

Reference to juridical consequences is clearly a universalized justification. Does this imply that behavioural consequences can not figure in legal justifications? It seems that if universalizability means that the reason given for a decision should apply not only to the particular case under scrutiny but also to any case which is similar in relevant aspects, then not only juridical but also behavioural, i.e. general social consequences are able to provide such a justification. The arguments based on general social consequences explicitly show their universalizable form: the decision-maker is concerned with the consequences of applying the favoured rule-interpretation in all similar future cases. In fact, it has been argued in the jurisprudential literature for quite some time that while universalizability is indeed a critical aspect of judicial decisions, "this requirement is quite consistent with reliance on policy".[47]

Finally, it should be noted that there is no necessary logical connection between consequence-based reasoning in law and consequentialism as a substantive moral standpoint. They neither implicate nor exclude one another.[48] It is not incoherent to reject pure consequentialism as a moral standpoint and to argue nevertheless that

---

[45]Note, 'Dworkin's "Rights Thesis"'; Greenawalt, 'Policy, Rights, and Judicial Decision'; MacCormick, 'Decisions', pp. 243 f.

[46]MacCormick, 'Decisions', p. 245, interpreting Dworkin.

[47]Greenawalt, p. 1010, n. 51.

[48]Cf. Barnett, 'Foreword: Of Chickens and Eggs – The Compatibility of Moral Rights and Consequentialist Analyses'; Cane, p. 43; Sinnott-Armstrong, 'Consequentialism'; Hooker, 'Rule Consequentialism'.

judges should justify at least some of their decisions in terms of certain consequences. Vice versa, it is a defensible (not self-contradictory) position for a legal philosopher to be a rule-utilitarian and to think that judges should not be allowed to justify their decisions in terms of their social effects.

Orthodox *law and economics* scholars once argued vigorously that common law judges should and in effect do, decide cases such as to maximize social welfare or "wealth".[49] This so-called "efficiency theory of the common law" is a *par excellence* consequentialist position both in the sense that it requires judges to base their decisions on consequences, namely their effect on social welfare, and in the sense that it is backed by a consequentialist moral theory, namely wealth maximization. However, from a logical point of view this link is contingent, i.e. purely accidental.

In summary, there does not seem to be a compelling philosophical argument for rejecting consequence-based judicial reasoning *tout court* as inconceivable or evidently unsound. A limited role of such arguments is compatible with various philosophical standpoints. As there are standard and uncontroversial cases when judges openly justify their decisions with consequence-based arguments,[50] it seems safe to say that consequence-based reasoning is conceivable. In the next sections I will argue that these arguments are also compatible with the institutional role of a judge in modern Western legal systems.

7 Feasibility: Objections from Individual and Collective Expertise

A consequence-based judicial decision can be represented as a three-step procedure of optimization under uncertainty. Ideally, a fully-informed rational decision-maker can solve this problem in an optimal way. First, identify the relevant normative standard(s), second, measure the consequences of each possible decision in the dimensions indicated by the standard(s), third, weight and compare the possible decisions and choose the one with the highest overall value. At each step, real-world courts run into difficulties. The objection from feasibility states that these difficulties are so serious that they make such an exercise in rational choice, and thus consequence-based reasoning impossible. Let us look at this argument more closely.

---

[49]See, e.g. Posner, *The Economics of Justice*. On the normative claim see Kornhauser, 'Guide'; 'Wealth'. On the positive claim see the articles in Rubin, *The Evolution of Efficient Common Law*. For a Rawlsian argument for an efficiency-oriented corporate law see Farber, 'Economic Efficiency and the Ex Ante Perspective'.

[50]See Wälde, *Juristische Folgenorientierung*; Coles, *Folgenorientierung im richterlichen Entscheidungsprozess*; Deckert, *Folgenorientierung in der Rechtsanwendung*.

Step	Question to be answered by the decision-maker	Difficulties
1. Identification	Which consequences (effects) matter?	Operationalization
2. Measurement	What is the impact of the decision in these dimensions?	Information
3. Evaluation	Which decision has better consequences overall?	Trade-offs

First, the judge has to identify which consequences of her decision are relevant. Some of these effects are easy to identify or even quantify, at least in theory. Others are notoriously difficult to operationalize. For instance, what would be the measure of a decision's impact on such legal values as predictability, legal certainty or coherence?[51]

In a second step, the judge has to measure the impact of her decision in all dimensions identified and operationalized in step one. Here she faces severe information imperfections and fundamental uncertainty about certain relevant variables. Factual information is almost always incomplete and not always reliable or verifiable. A further difficulty arises from the institutional setting of adjudication. In civil procedures as well as in the adversarial type of criminal procedure the informational base of the judge is essentially determined by the interested parties who do not have strong incentives to present their information in an unbiased way. Judges may rely on expert witnesses but this only provides a partial improvement. In addition to information problems, judicial procedures are also characterized by scarcity of resources such as time and human capital. Furthermore, the judges' cognitive limitations and lack of expertise in technical disciplines imply that even available information is not processed in a systematic and theoretically sound way.

Third, when choosing between alternatives, the judge has to evaluate the overall consequences of possible decisions in light of the relevant normative standards. If the relevant consequences cannot be easily reduced to one dimension, this choice involves difficult trade-offs. For example, a judge may have to answer questions like the following: how much legal certainty should be sacrificed in order to promote gender equality to a given extent?

In view of these difficulties, concerns about feasibility can be translated into the following question: what is likely to happen if a real-world judge has a duty to carry out such an optimization exercise, i.e. to assess the general social consequences of their decision in the way rational choice theory would suggest? Undoubtedly, they would face a plethora of problems. The pessimistic conclusion would be that an across-the-board mandate for consequence-based reasoning is likely to bring about intuitive, speculative or subjective decisions, eventually disguised as objective and well-founded. If decision-makers are authorized to base their reasoning on consequences but they lack information and expertise, such a mandate could backfire. Judges would enter into speculations about the behavioural consequences of their

[51] See, e.g. Hensche, pp. 105 ff.

decisions without any serious reliance on empirical evidence.[52] One critic comes to this conclusion:

> To the extent that sound empirical support is lacking for arguments about the likely impact of legal rules on human behaviour (i.e. we are ignorant about the likely behavioural consequences of legal rules), we need to develop criteria of good decision-making which do not depend upon knowledge of likely consequences.[53]

Is there a good reason to completely reject consequence-based reasoning as infeasible? Empirical research on judicial behaviour suggests that in case of (radically) insufficient information, time and technical expertise, judicial decisions are often still consequence-based. However, instead of solving a full-blown stochastic optimization problem, judges rely on heuristics and "rules of thumb".[54] In non-judicial contexts people make similarly complex decisions every day with tolerable results. At least in those domains of life where they are left free not to follow rules in a mechanical way, people adopt simple decision procedures, routines, "rules of thumb" and rely on heuristics which reduce complex decision-making problems into simple ones.

Do heuristics solve the problems faced by judges? Research suggests that "intuitive experts" reach results which are often tolerably close to the theoretical optimum.[55] While most of such mechanisms operate subconsciously, judges have to justify their decisions with reasonable public arguments. In contrast to non-legal contexts where intuitive decisions usually do not need further justification, adjudication remains in the domain of discursive rationality. It has been suggested that this public justification is not only required by political and moral principles such as the rule of law, but also leads to decisions which are better in substantive terms.[56] Reasoning can be made more transparent and accountable and in this sense more reasonable. The process of thinking about consequences and collecting necessary information has in itself a debiasing effect[57] when the brain switches from the heuristic to the calculative mode and the decision maker is compelled to provide plausible arguments in favour of her decision.

---

[52] Cane, p. 45. Cane provides the example of English law lords referring to potential over-deterrent effects of an extensive liability for medical malpractice without any actual reference to empirical data. Certainly, the point here is not whether such an effect could be demonstrated or made plausible. Rather the question is whether judges should be allowed to justify their decisions in a purely intuitive way, with no reference to empirical evidence.

[53] Cane, p. 43.

[54] See, e.g. Dhami, 'Psychological Models of Professional Decision Making'. For a systematic overview of the role of heuristics in law see Gigerenzer and Engel, *Heuristics and the Law*.

[55] On 'intuitive experts' see the results of the MPI Research Group led by Andreas Glöckner, http://www.coll.mpg.de/intuitiveexperts/max-planck-research-group-intuitive-experts, accessed on 4 April 2011.

[56] For a thorough analysis see Engel, 'The Impact of Representation Norms on the Quality of Judicial Decisions'.

[57] On debiasing see Jolls and Sunstein, 'Debiasing Through Law'.

Although it seems clear that consequence-based judicial reasoning faces serious problems, it is not clear what would be the reasonable "criteria of good decision-making which do not depend upon knowledge of likely consequences".[58] It is especially unclear why judges who would be required to follow these non-consequentialist criteria would rely less on heuristics. Before concluding that in real-world settings consequence-based reasoning, even when performed by conscientious judges, is infeasible or unavoidably leads to catastrophe, it has to be compared with feasible alternatives.[59] A reasonable comparison is with the competence of the legislator and administrative agencies in performing consequence-based decisions.

8 The Alternatives of Judicial Optimization: *Ex ante* Evaluation and Policy-Making in Legislation and Administration

Compared to the judiciary, the legislator and administrative agencies are better endowed, especially as far as technical expertise, resources and information are concerned.[60] On the other hand, legislators and administrative agencies are not necessarily more receptive towards non-partisan expertise. More often than not, or at least more often than the judiciary, they are subject to capture by well-organized private interests.

This would suggest that *ceteris paribus*, making policy decisions less political and more technocratic would lead to an improvement. In the last decades, many developed countries introduced systematic "*ex ante* evaluation" or "regulatory impact assessment", i.e. public and often mandatory consequence-based control of their legislative and administrative measures.[61] The judicial branch seems to be subject to entirely different standards. However, if we look at the rationales of *ex ante* evaluation, we can observe that most of them are relevant for adjudication as well.

Following Larouche we can distinguish the following rationales.[62] (1) *Ex ante* evaluation is a mechanism for collection of evidence. (2) It improves the quality of decision-making. Seen as a tool for improving quality, it follows an expert and technocratic logic: it purports to be objective, detached and possibly scientific; in the longer run, it contributes to an administrative or evaluation-oriented culture. (3) It

---

[58] Cane, p. 43.

[59] Cf. Komesar, *Imperfect Alternatives*.

[60] It can even be argued that this should remain the case. As the expertise in assessing the systemic consequences of legal rules, i.e. the human capital of those who are knowledgeable in both legal and technical or social scientific questions, is a scarce resource, this human capital is more effectively used in legislative drafting *ex ante* than in case-by-case interpretative corrections *ex post*. Cf. Eidenmüller, 'Rechtswissenschaft als Realwissenschaft'.

[61] Pfaff and Guzelian, 'Evidence Based Policy'; Verschuuren, *The Impact of Legislation*.

[62] Larouche, '*Ex ante* evaluation of legislation torn among its rationales'.

increases transparency and openness; (4) it makes decision-making more democratic by allowing for participation of stakeholders; (5) it contributes to justification: the process of conducting an *ex ante* evaluation is a way to explain publicly why the proposed action is necessary and appropriate; (6) it also increases accountability by highlighting the trade-offs being made by the decision-maker. While the first two rationales are result oriented, the last four are process oriented.

From a conventional view of the judicial role, *ex ante* evaluation looks strange and inappropriate. However, an *ex ante* evaluation followed by the decision of a democratically legitimized body (legislation) and a court decision supported by a consequence-based justification are very similar. What happens before the decision in the two cases can be the same in a technical sense. On closer inspection, the result oriented rationales seem to apply to those court decisions which have large-scale consequences, for example in the case of highest court rulings. As for the procedural values, with the potential exception of (4), they seem to be equally relevant. While judicial decisions are not expected to be democratic neither in representative, participatory or deliberative sense, it is difficult to argue that the other rationales, especially transparency and accountability are not desirable for judicial decision-making.

Ideally, a fully informed rational decision-maker can solve the problem of consequence-based decision-making in an optimal way. However, real-world decision-makers run into serious difficulties. In legislation and to an even larger extent in adjudication, time constraints are strict, resources, including expertise, are limited and the use of reliable control mechanisms such as experimentation is rare. On the other hand "the law" as a practice cannot be suspended until the best theoretical solutions are found or all the relevant consequences of a decision are carefully examined. In consequence, pragmatism and "simple rules"[63] are used: decisions are based on various heuristics and "rules of thumb". As the use of these heuristics is not always conscious and self-reflexive, competence also raises institutional and normative problems of transparency, justification, accountability or, in summary, legitimacy. This leads us to the third objection against consequence-based arguments: its desirability.

9 Desirability and Legitimacy

When it comes to the desirability of consequence-based reasoning, the black-letter lawyer would ask primarily doctrinal questions, usually with reference to a particular legal area and/or procedure, as to which consequences under what conditions should be allowed to have what weight in the judicial reasoning. In contrast, the legislator, the social critic and the informed public would be interested in pragmatic or political questions related to particular policy issues or broader problems of institutional design. Their question would be: how much role should be assigned to judges

[63]Cf. Epstein, *Simple Rules for a Complex World.*

for openly evaluating the effects of their decisions. Both kind of questions can refer to various legal issues, ranging from the detention of a criminal suspect, through the prohibition of anticompetitive commercial practices to the interpretation of an agreement for surrogate motherhood.

As the previous sections suggested, consequence-based reasoning is not necessarily beyond the expertise of courts. This does not mean that the way judges currently conduct it is satisfactory. The imperfections of judicial decision-making and the biases in judicial reasoning bear normative relevance when it comes to (small-scale, marginal) legal reforms or (large-scale, total) institutional design.[64] If judges face serious and systematic difficulties in performing their duties in the way their role would require them to do, then these role-defined duties or normative expectations should be revised and eventually modified. If it is established that decision-makers commit systematic errors in evaluating certain types of evidence (for instance, they fall prey to the hindsight bias[65]), this gives reason for a benevolent institutional designer to change the rules governing legal procedures. An ideal institutional designer would take the heuristics and biases of legal decision-makers into account by both harvesting their beneficial effects, if any, and compensating the detrimental ones.

It should also be kept in mind that in real-world settings, judges who have a predominantly legal training are not in a position to undertake complex probability calculations or full-blown statistical analyses. To some extent, their work can be improved by decision support systems and artificial intelligence.[66] This does not mean that consequence-based decisions are merely technical, neutral or "objective" in the sense of being merely factual. The expected consequences have to be evaluated and whether or not this is called "balancing", value-laden trade-offs have to be made.[67]

As stressed by the regulatory perspective, modern law has to fulfil conflicting normative expectations by combining doctrinal coherence with effectiveness and responsiveness.[68] One particular concern about consequence-based reasoning seems to be whether the consequentialism inherent in effectiveness and responsiveness inevitably corrupts law's non-instrumental commitments to doctrines and principles associated with values such as coherence and procedural justice.[69] This question is closely related to the institutional role of courts and judges – in this case not to the nature or "essence" of judging, rather to the normative question, what are the tasks judges should do, considering what they are able to do? Here the answer will be probably domain specific again. As the "point", "policy purpose" or "function" of

[64]On the normative relevance of behavioural research, see Tor, pp. 298 ff.

[65]Rachlinski, 'A Positive Psychological Theory of Judging in Hindsight'.

[66]Bell, 'Policy', p. 73.

[67]Koch and Rüßmann, p. 229.

[68]Parker et al., 'Introduction'.

[69]Parker et al., p. 11. Although the authors also mention distributive justice as conflicting with consequentialism, this is an evidently consequence-based consideration.

legal institutions differ, we might want to share the competence for forward-looking decisions between legislation (rule-setting) and adjudication (rule-application) differently in case of e.g. car accidents, patent protection or freedom of speech. The balance is likely to be highly dependent on the particular context.

Finally, one should note that the three objections from conceivability, desirability and feasibility, and the answers to them are interrelated. From the perspective of institutional design, the "feasible" sets limits to the "desirable" and all this has to be within the universe of the "conceivable." I have argued that the most serious objections relate to feasibility. Feasibility in itself is not a normative but an empirical matter – but to a large extent it depends not on brute but on institutional facts.[70] Various legal cultures differ largely in their institutional variables, their conventional understanding of the judicial role, the canon of accepted arguments and many other dimensions which determine the action space and the argumentative tools at the disposal of judges.

10 Conclusions

While standard *law and economics* research usually focuses on the outcome of legal cases or, less frequently, on the motivational determinants and institutional context of judicial behaviour,[71] *law and economics* scholars should also confront problems of legal reasoning. This would both contribute to the jurisprudential self-consciousness of the discipline and improve the chances for economic arguments to have an impact on important legal decisions.

The efficiency-based policy recommendations as to how judges should interpret rules and decide cases are relevant for judges to the extent that prudential or consequence-based arguments are relevant for the justification of legal decisions. Although *law and economics* scholars do not always make this proviso explicit, they should be aware of the methodological status of their arguments. Arguably, many of them accept this "jurisprudential preface" at least implicitly when they say that the practical impact of their findings is conditional upon certain characteristics of particular political communities or legal systems.

The analysis presented here also has a further lesson for *law and economics* scholarship, especially for the discussion on "legal origins".[72] If we are interested in how law contributes to the welfare of society, part of the question will concern the mechanisms through which adjudication determines the quality of law. As Gillian Hadfield argued in a recent paper, "the quality of a legal regime is in part a function of its capacity to adapt to local and changing circumstances."[73] In particular,

[70]Cf. Searle, *Speech Acts*.

[71]Arrunada and Andonova, 'Judges' Cognition and Market Order'; Kornhauser, 'Modelling Courts'; Posner, *How Judges Think?*

[72]La Porta et al., 'The Economic Consequences of Legal Origins'.

[73]Hadfield, p. 45.

she tried to identify the "conditions under which the law (judges) will learn [new welfare-relevant] information about the environment and incorporate that information into the [...] rule."[74] A fundamental question in this respect is whether this adaptation occurs differently in legal systems which give judges discretion and cultivate a judicial ideology of taking social consequences into account than in those legal systems where the canon binds judges more closely to rule-based arguments. Arguably, the latter have more subtle and indirect ways of adaptation. This issue shows that consequence-based reasoning might be relevant for *law and economics* in yet another way, as a variable in a comparative analysis of the determinants of the quality of law, as an institutional determinant of economic performance.

Bibliography

Ackerman, Bruce A., 'Law, Economics, and the Problem of Legal Culture', in *Duke Law Journal* (1986), pp. 929 ff.

Anderson, Bruce, 'Context of Discovery, Context of Decision and Context of Justification in the Law', in *IVR encyclopedia* (2009), http://ivr-enc.info/index.php?title=Context_of_Discovery,_Context_of_Decision_and_Context_of_Justification_in_the_Law (accessed on 4 April 2011)

Arrunada, Benito and Andonova, Veneta, 'Judges' Cognition and Market Order', in *Review of Law and Economics*, Vol. 4 (2008), pp. 665 ff.

Barnett, Randy, 'Foreword: Of Chickens and Eggs – The Compatibility of Moral Rights and Consequentialist Analyses', in *Harvard Journal of Law and Public Policy*, Vol. 12 (1989), pp. 611 ff.

Bell, John, 'The Acceptability of Legal Arguments', in Neil D. MacCormick and Peter Birks (eds.), *The Legal Mind. Essays for Tony Honoré* (Oxford, 1986), pp. 45 ff. (cited as: 'Acceptability')

Bell, John, 'Policy Arguments and Legal Reasoning', in Zenon Bankowski et al. (eds.), *Informatics and the Foundations of Legal Reasoning* (Dordrecht, 1995), pp. 73 ff. (cited as: 'Policy')

Breyer, Stephen, *Economic Reasoning and Judicial Review* (Washington, 2004)

Burton, Steven J., 'Comment on Professor Patterson's *Pseudo-Debate over Default Rules in Contract Law*', in *Southern California Interdisciplinary Law Journal*, Vol. 3 (1993), pp. 303 ff.

Cane, Peter, 'Consequences in Judicial Reasoning', in Jeremy Horder (ed.), *Oxford Essays in Jurisprudence*, Fourth Series (Oxford, 2000), pp. 41 ff.

Coles, Christina, *Folgenorientierung im richterlichen Entscheidungsprozess. Ein interdisziplinärer Ansatz* (Frankfurt a.M., 1991)

Craswell, Richard, 'Default Rules, Efficiency, and Prudence', in *Southern California Interdisciplinary Law Journal*, Vol. 3 (1993), pp. 289 ff.

Cserne, Péter, 'Policy Arguments Before Courts: Identifying and Evaluating Consequence-Based Judicial Reasoning', in *Humanitas Journal of European Studies*, Vol. 3 (2009), pp. 9 ff.

Dau-Schmidt, Kenneth G., and Carmen L. Brun, 'Lost In Translation: The Economic Analysis of Law in the United States and Europe', in *Columbia Journal of Transnational Law*, Vol. 44 (2006), pp. 602 ff.

Deckert, Martina Renate, *Folgenorientierung in der Rechtsanwendung* (Munich, 1995)

Dhami, Mandeep K., 'Psychological Models of Professional Decision Making', in *Psychological Science*, Vol. 14 (2003), pp. 175 ff.

Dworkin, Ronald, *Taking Rights Seriously* (Cambridge MA, 1977)

Eidenmüller, Horst, 'Rechtswissenschaft als Realwissenschaft', in *Juristenzeitung* 54 (1999), pp. 53 ff.

[74]Hadfield, p. 46.

Engel, Christoph, 'The Impact of Representation Norms on the Quality of Judicial Decisions', in *MPI Collective Goods Preprints*, No. 2004/13, available at SSRN: http://ssrn.com/abstract=617821

Epstein, Richard A., *Simple Rules for a Complex World* (Cambridge MA, 1995)

Farber, Daniel A., 'Economic Efficiency and the Ex Ante Perspective', in Jody S. Kraus and Steven D. Walt (eds.), *The Jurisprudential Foundations of Corporate and Commercial Law* (Cambridge MA, 2000), pp. 54 ff.

Garoupa, Nuno and Thomas S. Ulen, 'The Market for Legal Innovation: Law and Economics in Europe and the United States', in *Alabama Law Review*, Vol. 59 (2008), pp. 1555 ff.

Gazal-Ayal, Oren, 'Economic Analysis of Law in North America, Europe and Israel', in *Review of Law & Economics*, Vol. 3/2 (2007), Article 11

Gigerenzer, Gerd and Engel, Christoph (eds.), *Heuristics and the Law* (Boston, 2006)

Grechenig, Kristoffel and Gelter, Martin, 'The Transatlantic Divergence in Legal Thought: American Law and Economics vs. German Doctrinalism', in *Hastings International and Comparative Law Review*, Vol. 31 (2008), pp. 295 ff.

Greenawalt, Kent, 'Policy, Rights, and Judicial Decision', in *Georgia Law Review*, Vol. 11 (1977), pp. 991 ff.

Hadfield, Gilian K., 'The Levers of Legal Design: Institutional Determinants of the Quality of Law', in *Journal of Comparative Economics*, Vol. 36 (2008), pp. 43 ff.

Hage, Jaap, 'Legal Reasoning', in Jan Smits (ed.), *Elgar Encyclopedia of Comparative Law* (Cheltenham, 2006)

Harris, Ron, 'The Uses of History in Law and Economics', in *Theoretical Inquiries in Law*, Vol. 4 (2003), pp. 659 ff.

Hensche, Martin, 'Probleme einer folgenorientierten Rechtsanwendung', in *Rechtstheorie*, Vol. 29 (1998), pp. 103 ff.

Honoré, A. M., 'Legal Reasoning in Rome and Today', in *Cambrian Law Review*, Vol. 4 (1973), pp. 58 ff.

Hooker, Brad, 'Rule Consequentialism', in *Stanford Encyclopedia of Philosophy* (2008), available at: http://plato.stanford.edu/entries/consequentialism-rule/

Jolls, Christine and Sunstein, Cass R., 'Debiasing Through Law', in *Journal of Legal Studies*, Vol. 35 (2006), pp. 199 ff.

Kennedy, Duncan, 'Legal Formality', in *Journal of Legal Studies*, Vol. 2 (1973), pp. 351 ff. (cited as: 'Formality')

Kennedy, Duncan, 'Form and Substance in Private Law Adjudication', in *Harvard Law Review*, Vol. 89 (1976), pp. 1685 ff. (cited as: 'Form')

Kirchner, Christian, 'Zur konsequentialistischen Interpretationsmethode: Der Beitrag der Rechtswissenschaft zur reziproken methodischen Annäherung von Ökonomik und Rechtswissenschaft', in Thomas Eger et al. (eds.), *Internationalisierung des Rechts und seine ökonomische Analyse. Festschrift für Hans-Bernd Schäfer zum 65. Geburtstag* (Wiesbaden 2008), pp. 37 ff.

Koch, Hans-Joachim and Rüßmann, Helmut, *Juristische Begründungslehre. Eine Einführung in Grundprobleme der Rechtswissenschaft* (Munich, 1982)

Komesar, Neil K., *Imperfect Alternatives. Choosing Institutions in Law, Economics, and Public Policy* (Chicago, 1994)

Kornhauser, Lewis A., 'A Guide to the Perplexed Claims of Efficiency in the Law', in *Hofstra Law Review*, Vol. 8 (1980), pp. 591 ff. (cited as: 'Guide')

Kornhauser, Lewis A., 'Wealth Maximization', in Peter Newman (ed.), *The New Palgrave Dictionary of Economics and the Law* Vol. 3 (London, 1998), pp. 679 ff. (cited as: 'Wealth')

Kornhauser, Lewis A., 'Modelling Courts', in Mark D. White (ed.), *Theoretical Foundations of Law and Economics* (Cambridge, 2008), pp. 1 ff.

La Porta, Rafael et al., 'The Economic Consequences of Legal Origins', in *Journal of Economic Literature*, Vol. 46 (2008), pp. 285 ff.

Larouche, Pierre, '*Ex Ante* Evaluation of Legislation Torn Among its Rationales', in Jonathan M. Verschuuren (ed.), *The Impact of Legislation. A Critical Analysis of* Ex Ante *Evaluation* (Leiden, 2009), pp. 39 ff.

Luhmann, Niklas, *Rechtsdogmatik und Rechtssystem* (Stuttgart, 1974; cited as: *Rechtsdogmatik*)

Luhmann, Niklas, 'Legal Argumentation: An Analysis of Its Form', in *Modern Law Review*, Vol. 58 (1995), pp. 285 ff. (cited as: 'Legal Argumentation')

Luhmann, Niklas, *Risk. A Sociological Theory*, trans. Rhodes Barrett (New Brunswick, 2005; cited as: *Risk*)

Lübbe-Wolff, Gertrude, *Rechtsfolgen und Realfolgen. Welche Rolle können Folgenerwägungen in der juristischen Regel- und Begriffsbildung spielen?* (Freiburg, 1981)

MacCormick, Neil, *Legal Reasoning and Legal Theory* (Oxford, 1978; cited as: *Reasoning*)

MacCormick, Neil, 'On Legal Decisions and Their Consequences: From Dewey to Dworkin', in *New York University Law Review*, Vol. 58 (1983), pp. 239 ff. (cited as: 'Decisions')

MacCormick, Neil, *Rhetoric and the Rule of Law* (Oxford, 2005; cited as: *Rhetoric*)

MacCormick, Neil and Summers, Robert S., *Interpreting Statutes. A Comparative Study* (Aldershot, 1991)

Note, 'Dworkin's "Rights Thesis"', in *Michigan Law Review*, Vol. 74 (1976), pp. 1167 ff.

Parker, Christine et al., 'Introduction', in Parker et al. (eds.), *Regulating Law* (Oxford, 2004), pp. 1 ff.

Patterson, Dennis, 'The Pseudo-Debate over Default Rules in Contract Law', in *Southern California Interdisciplinary Law Journal*, Vol. 3 (1993), pp. 235 ff.

Pfaff, John F. and Guzelian, Christopher P., 'Evidence Based Policy', in *Fordham Law Legal Studies Research Paper* No. 976376 (2007), available at SSRN: http://ssrn.com/abstract=976376

Posner, Richard A., *The Economics of Justice* (Cambridge MA, 1981)

Posner, Richard A., 'The Future of the Law and Economics Movement in Europe', in *International Review of Law and Economics*, Vol. 17 (1997), pp. 3 ff.

Posner, Richard A., *How Judges Think?* (Cambridge MA, 2008)

Rachlinski, Jeffrey J., 'A Positive Psychological Theory of Judging in Hindsight', in Cass R. Sunstein (ed.), *Behavioral Law and Economics* (Cambridge MA, 2000), pp. 95 ff.

Rubin, Paul H. (ed.), *The Evolution of Efficient Common Law* (Cheltenham, 2007)

Rudden, Bernard, 'Consequences', in *Juridical Review*, Vol. 24 (1979), pp. 193 ff.

Sajó, András, 'How the Rule of Law Killed Hungarian Welfare Reform', in *East European Constitutional Review*, Vol. 5/1 (1996), pp. 31 ff.

Schauer, Frederick, *Playing by the Rules. A Philosophical Examination of Rule- Based Decision-Making in Law and in Life* (Oxford, 1991)

Schiemann, Gregor, 'Inductive Justification and Discovery. On Hans Reichenbach's Foundation of the Autonomy of the Philosophy of Science', in Jutta Schickore and Friedrich Steinle (eds.), *Revisiting Discovery and Justification: Historical and Philosophical Perspectives on the Context Distinction* (Dordrecht, 2006), pp. 23 ff.

Searle, John, *Speech Acts. An Essay in the Philosophy of Language* (Cambridge, 1969)

Sibony, Anne-Lise, *Le juge et le raisonnement économique en droit de la concurrence* (Paris, 2008)

Sinnott-Armstrong, Walter, 'Consequentialism', in *Stanford Encyclopedia of Philosophy* (2006), available at: http://plato.stanford.edu/entries/consequentialism/

Symposium 'Economic Analysis in Civil Law Countries: Past, Present, Future', in *International Review of Law and Economics*, Vol. 11 (1991), pp. 261 ff.

Teubner, Gunther, 'Folgenkontrolle und responsive Dogmatik', in *Rechtstheorie*, Vol. 6 (1975), pp. 179 ff. (cited as: 'Folgenkontrolle')

Teubner, Gunther (Hrsg.), *Entscheidungsfolgen als Rechtsgründe. Folgenorientiertes Argumentieren in rechtsvergleichender Sicht* (Nomos, 1995; cited as: *Entscheidungsfolgen*)

Tor, Avishalom, 'The Methodology of the Behavioural Analysis of Law', in *Haifa Law Review*, Vol. 4 (2008), pp. 237 ff.

Ulen, Thomas S., 'Courts, Legislatures, and the General Theory of Second Best in Law and Economics', in *Chicago-Kent Law Review*, Vol. 73 (1998), pp. 189 ff.

Verschuuren, Jonathan M. (ed.), *The Impact of Legislation. A Critical Analysis of* Ex Ante *Evaluation* (Leiden, 2009)

Vogenauer, Stefan, *Die Auslegung von Gesetzen in England und auf dem Kontinent* (Tübingen, 2000)

Wald, Patricia M., 'Limits on the Use of Economic Analysis in Judicial Decision-making', in *Law and Contemporary Problems*, Vol. 50 (1987), pp. 225 ff.

Wälde, Thomas W., *Juristische Folgenorientierung. "Policy analysis" und Sozialkybernetik. Methodische und organisatorische Überlegungen zur Bewältigung der Folgenorientierung im Rechtssystem* (Königstein, 1979)

Is the Rationality of Judicial Judgements Jeopardized by Cognitive Biases and Empathy?

Klaus Mathis and Fabian Diriwächter

The introduction to this essay embarks on a discussion of the extent to which cognitive biases jeopardize the rationality of judicial judgements and, consequently, which interventions or institutional precautions are necessary to ensure that the judgements delivered by judges are nevertheless as rational as possible. The next question to be considered is whether a judge should necessarily possess empathy, or whether this might not undermine impartiality and ultimately, in a similar way as cognitive biases, jeopardize the rationality of court judgements. The results of these two lines of inquiry are summed up in the conclusion.

1 Introduction

The phenomenon of cognitive biases has long been investigated within the field of decision psychology. The "behavioural economics" branch of economic theory has made use of the resulting insights by elucidating the psychological foundations of the economic theory of decision-making, or "rational choice", in order to enhance the explanatory power of economic models. The specific matter at issue is the assumption of rationality, which presumes individuals on principle to have the capacity to act in their own self-interest, i.e. to assess and evaluate their scope for action in such a way as to maximize their own utility.

Of course, it was observed early on that the individual is not like an omniscient computer, ever ready to calculate the best of all possible options at lightning speed.[1] Consequently the concept of "bounded rationality" developed by Herbert A. Simon is rooted in the knowledge that human cognitive capabilities are limited.[2] An important milestone in this regard is the "prospect theory" of Daniel Kahneman and Amos

[1] Mathis, p. 25.

[2] See Herbert A. Simon, 'Rational Decision Making in Business Organizations'.

K. Mathis (✉)
University of Lucerne, Faculty of Law, Frohburgstrasse 3, P.O. Box 4466, CH-6002 Lucerne, Switzerland
e-mail: klaus.mathis@unilu.ch

K. Mathis (ed.), *Efficiency, Sustainability, and Justice to Future Generations*, Law and Philosophy Library 98, DOI 10.1007/978-94-007-1869-2_3,

Tversky, in which Simon's work is taken forward and the assumption of rationality is further relativized.[3]

Building upon these insights, Kahneman and Tversky also studied the phenomenon of cognitive biases, which will be analysed more closely in the following with a focus on judicial decision-making. To illustrate this problematique, the issues are exemplified principally with reference to the assessment of negligence in tort law. The discussion then proceeds to consider how and to what extent cognitive biases can be avoided or their impact at least mitigated or compensated. The final question raised is whether a judge ought to have empathy, and whether this does not affect judicial impartiality and thus, like cognitive biases, ultimately jeopardize the rationality of judicial decision-making.

2 Cognitive Biases

In their essay "Judgement under Uncertainty: Heuristics and Biases", published in 1974, Tversky and Kahnemann asserted that people in complex decision-making situations resort to what are known as mental heuristics. These fall into the category of automatic thought-processes which take place unconsciously, unintentionally, involuntarily and effortlessly. The use of such cognitive rules of thumb is fundamentally useful and maybe even essential in order to cope with complex situations in everyday life. Sometimes, however, it leads to serious systematic errors.[4] Just as distortions of visual perception can cause reality to be misinterpreted, mental heuristics can also potentially give rise to misjudgements.[5] These are commonly referred to as "biases".[6]

Even judges are prone to make cognitive misjudgements. The errors in question are not just occasional lapses, but symptoms of a problem that causes *systematic distortions of judgement*:

> The quality of the judicial system depends upon the quality of decisions that judges make. Even the most talented and dedicated judges surely commit occasional mistakes, but the public understandably expects judges to avoid systematic errors. This expectation, however, might be unrealistic.[7]

There are many diverse forms of cognitive biases which may distort judicial decision making.

---

[3]See Daniel Kahneman and Amos Tversky, 'Prospect Theory: An Analysis of Decision under Risk'.

[4]Tversky and Kahneman, p. 1124.

[5]Guthrie, Rachlinski and Wistrich, p. 780.

[6]Tversky and Kahneman, p. 1124.

[7]Guthrie, Rachlinski and Wistrich, p. 778.

2.1 Heuristics and Biases

The following section will begin with brief explanations of the most important heuristics and the associated cognitive distortions, illustrated with examples taken primarily from the assessment of negligence in tort law.

2.1.1 Availability Bias

The heuristics of availability mean that the *subjective probability* of an event is all the greater the more easily or quickly one is able to imagine or recall examples of the event:[8]

> [P]eople might resolve a question of probability not by investigating statistics, but by asking whether a relevant incident comes easily to mind.[9]

Thus it is possible that judging the probability of a traffic accident depends upon how vividly one can imagine such an event. Here, one's own experiences are the most influential factor. Somebody who has recently witnessed a serious road accident will consider the probability of such accidents to be considerably higher than someone who has not witnessed an accident for a very long time.[10] For instance, judges in the civil courts regularly have damages cases brought before them and are therefore able to call such incidents to mind with greater ease. Or to express it another way: for judges *as individuals*, accidents and incidents giving rise to damages in general are *more readily cognitively available*.

The factor of individual cognitive availability is complemented by the *general cognitive presence* of traffic accidents. This is an aspect in which the media play a key role. For instance, the circumstance that far more car-crash fatalities are reported in the news than fatalities from cancer encourages the impression that more people die of the consequences of a traffic accident than of cancer, which is clearly incorrect.[11]

Another relevant factor is known as *illusory correlation* (the mental linking of unrelated events). Because people clearly pay attention to positive contingencies, they overestimate the frequency of the simultaneous occurrence of certain events. The alleged correlation between hashish use and delinquency can be cited as an example. Here, the negative contingencies are often neglected or simply ignored. In point of fact, people may smoke hashish without ever having committed an offence; or they may commit offences without ever having smoked hashish; while the third and final possibility is that a person has never committed an offence, nor ever smoked hashish.[12]

---

[8]Tversky and Kahneman, p. 1127.

[9]Jolls, p. 77.

[10]Jungermann, Pfister and Fischer, p. 173.

[11]Jungermann, Pfister and Fischer, p. 174.

[12]Jungermann, Pfister and Fischer, p. 174.

The following case provides a useful illustration of the heuristic of availability in judicial judgements: two motorcars collide. One of the drivers, a 50-year-old in a people-carrier, states that the collision was caused by the other vehicle's excessive speed. If the other driver is a "boy racer" with a souped-up sports car, what will be his prospects in court if the skid-marks do indeed suggest that he was driving a little too fast? The illusory correlation is obvious: boy racers with souped-up sports cars regularly drive too fast. This thesis is "confirmed" by the media reporting of high-speed crashes. Nevertheless, it is equally possible that, in the case described, the driver of the people-carrier also committed some transgression. Perhaps he was distracted by taking a mobile phone call at the wheel or because he was concentrating on tuning the car radio. It is conceivable that the boy racer's chances of a favourable judgement will be poor unless he can present evidence of the opposing party's misdemeanour, which is almost impossible.

2.1.2 Hindsight Bias

Like the truism that one is always wiser after the event, many events, particularly those with a tragic outcome, seem predictable in retrospect. Baruch Fischhoff was the first to address this phenomenon, which has since been named "hindsight bias". The question he asked was how knowledge of the outcome of a sequence of events affects its *ex post* assessment. Over the course of several experiments, he substantiated his initial hypothesis of retrospective overestimation of the predictability of an event.[13] Knowledge of the outcome of a sequence of events influences people's *ex post* assessments insofar as they believe that people should have had better foresight than was actually possible. This phenomenon can also be described as the "knew-it-all-along effect" or "creeping determinism".[14]

The fact that those forming a judgement in retrospect tend to *overestimate the predictability of events* has since been confirmed in numerous other studies.[15] Drawing on Fischhoff, Jeffrey J. Rachlinski tried to explain this: in his view, people wanted to assimilate the outcome of a sequence of events and the circumstances that led to this actual consequence into a coherent whole. They are thought to do so as follows: pieces of information that seem relevant in the light of the outcome are treated as especially relevant from the very outset. At the same time the importance of information which, *ex ante*, might have indicated an alternative course of events, is played down.[16]

This cognitive error, according to Rachlinski, is further amplified by *motivational factors*: many people yearn for stability and predictability. And since the predictability of a disastrous event implies its avoidability, the hindsight bias may

[13]Fischhoff, p. 292.

[14]Fischhoff, p. 288.

[15]See Jay J. Christensen-Szalanksi and Cynthia Fobian-Willham, 'The Hindsight Bias: A Meta-Analysis'.

[16]Rachlinski, p. 97.

create an incredibly soothing effect by making events appear controllable.[17] Even if an event, considered *ex ante*, may appear to be completely singular and therefore unpredictable, and with sufficient effort of will this could also be confirmed *ex post*, it seems more comforting – particularly in the face of the tragic fates of individuals – to rank that sequence of events as part of the normal course of things, thereby rendering it ostensibly more manageable for the future.

The hindsight bias is – as Rachlinski states – of enormous importance for judicial practice:

> Because courts primarily judge in hindsight, the bias might exert tremendous influence on judgement in the legal system.[18]

This problematique will be illustrated with an applied example chosen for its significance to legal practice: the assessment of negligence in fault-based liability. The analysis must be prefaced with the remark that the Swiss law on non-contractual liability makes a distinction between fault-based liability and causal forms of liability. The latter are further differentiated into simple forms of causal liability, and liability for risk (also called strict or aggravated causal liability). Whereas the attributive criterion for fault-based liability is the accusation that the injurer failed to prevent an ostensibly foreseeable injury, the various categories of causal liability are defined by the existence of certain *standardized circumstances*, such as dereliction of the duty of diligence for simple causal liability, or the occurrence of a potential for injury associated with hazardous conditions in the case of risk liability.[19]

In the present context, the influence of hindsight bias on fault-based liability is of particular interest. In Switzerland, Article 41 para. 1 of the Swiss Code of Obligations (corresponding roughly to liability pursuant to Section 823 (1) of the German Civil Code) provides for this and imposes a fundamental obligation on the party at fault to pay compensation for damage caused unlawfully, whether wilfully or negligently, to another party. Cases of wilful damage are relatively rare in liability law; of greater significance by far is negligence liability.

According to the traditional view, *negligence* is understood to mean that a person of sound mind has not acted with the degree of diligence that the average reasonable person would have exercised in the same circumstances. Accordingly, negligence consists in deviation from the behaviour expected of the average person.[20] For someone to behave properly, they must not only be capable of assessing the consequences of their action but must also have the strength of will to prevent an anticipated potential occurrence of damage by taking due care.[21] The injurer, then, becomes liable if he could have anticipated that his actions might cause injury to a third party and could have avoided this by adhering to the due standard of care. If, on the other

[17] Rachlinski, pp. 97 f.

[18] Rachlinski, p. 99.

[19] On this whole area see Rey, N 58 ff.

[20] Rey, N 844.

[21] Rey, N 810.

hand, the damage is held to have been unforeseeable, in principle there is no basis on which to make the causal agent accountable.[22]

Foreseeability does not therefore mean that liability is limited to actually foreseen injuries – it is determined rather by what the injurer could or should have foreseen.[23] A negligent action is committed on the one hand by somebody who causes an injury due to lack of care because he does not foresee what a careful person in the same position would be able to foresee, and on the other hand by somebody who does foresee a possible injury but does not take sufficient care to prevent its occurrence.[24] The *probability of occurrence* of some infringement of a legally protected interest is relevant here insofar as it permits a conclusion to be drawn about what the average reasonable person would be able to foresee. Therefore, the greater the probability of occurrence, the more the insurer ought to have reckoned with the occurrence of the injury of legal interests, the more avoidable they would have been and the more compelling the allegation of fault.

The problematic issue here is that the judge cannot blank out the events which are now already known, and can consequently no longer make a wholly objective *ex ante* assessment: because things turned out the way they evidently had to, the injury's probability of occurrence is overestimated and the judge subject to hindsight bias believes that the injurer's foresight should have been better than was actually possible.

What is clear from these considerations is something that has long been criticized in legal theory: only rarely do the courts deny the foreseeability of the consequences of a certain behaviour. Instead, they demand an extremely high degree of foresight that no reasonable person can muster, and level the accusation of negligence with undue alacrity. One reason for this judicial stringency lies in hindsight bias.[25] And indeed, various studies which demonstrate the influence of hindsight bias on judgements in liability cases.[26]

Rachlinski found that due to hindsight bias, negligence liability was shifting ever closer to causal liability. He concluded that hindsight bias therefore caused the discrepancy between postulated law and the application of law.[27] An interesting parallel can be noted in the Swiss juristic literature, where fault-based liability is also shifting towards a general causal-agent liability, and a discrepancy has arisen between postulated and implemented law.[28]

[22]Oftinger and Stark, § 1 N 30 ff.

[23]Oftinger and Stark, § 5 N 69.

[24]Oftinger and Stark, § 5 N 48 ff.; Brehm, N 196 ff. on Art. 41, Swiss Code of Obligations.

[25]Falk, pp. 10 ff.; Schweizer pp. 224 ff.

[26]See Kim A. Kamin and Jeffrey J. Rachlinski, 'Ex Post ≠ Ex Ante. Determining Liability in Hindsight'; LaBine Susan J. and LaBine Gary, 'Determinations of Negligence and the Hindsight Bias'; Reid Hastie, David A. Schkade and John W. Payne, 'Juror Judgements in Civil Cases: Hindsight Effects on Judgements of Liability for Punitive Damages'.

[27]Rachlinski, pp. 99 f.

[28]Oftinger and Stark, § 5 N 81.

2.1.3 Anchoring

The concept of "anchoring" denotes the way that people often make probability judgements intuitively or with reference to entrenched views (anchors), which are often entirely *arbitrary* and consequently maladapted to the actual conditions.[29] Anchoring can result in both numerical misjudgements as well as memory distortions. In a study by Tversky and Kahneman, students were asked to estimate the proportion of African states in the United Nations Organization. Before the estimation task, the experimental subjects spun a wheel of fortune which "randomly" stopped at the number 65 or 10. Students in the group whose lucky number was 65 estimated the proportion of African states in the UN at 45% on average; for the group with the lucky number 10, the mean estimated value was 25%. The completely arbitrary value of the wheel of fortune did indeed influence the subjects' estimates – it acted as an anchor.[30] Even if people recognize the invalidity of an anchor, it serves as a reference point for their subsequent estimate because they start from the anchor and make an adjustment upward or downward.[31]

Anchoring can also lead to *memory distortions*, as the following example may illustrate: an experimental subject is asked to estimate the length of the Rhine. Because she does not know the exact length, she settles on the completely arbitrary figure of 1,050 km. A few weeks later, the test subject is informed of the true length of the Rhine (1,230 km) and asked to recall her own estimate. If the test subject can no longer recall her original estimate precisely, she considers what she might have estimated a week ago and comes up with a result of 1,150 km.[32] The memory, which is no longer secure, is adjusted to the information imparted immediately beforehand. That information is serving as an anchor.

A kind of anchoring effect may play a part in the assessment of negligence in fault-based liability. It is self-evident that in making a judgement on foreseeability, only the *ex ante* perspective, i.e. the perspective of the injurer at the time just prior to the occurrence of the injury, can be relevant, since the defendant is only deemed to be liable if he could have avoided its occurrence. However, knowledge about the outcome of the sequence of events now acts as a mental anchor for judging the predictability of this event.[33] The information that "damage occurred" can produce a distortion of almost 100% in the judge's judgement on the probability of occurrence, which in turn misleads him to draw a wrong conclusion about the actual foreseeability of the outcome. The inference that the defendant ought to have foreseen the sequence of events, could therefore have been in a position to avert damage, and is therefore at fault, is favoured by the anchoring effect.[34] In this way, anchors can also amplify the above-mentioned hindsight bias.

[29] Tversky and Kahneman, p. 1128; Jungermann, Pfister and Fischer, pp. 174 ff.

[30] Tversky and Kahneman, p. 1128.

[31] Guthrie, Rachlinski and Wistrich, p. 788.

[32] Cf. Jungermann, Pfister and Fischer, p. 175.

[33] Chapman and Johnson, p. 134.

[34] Cf. Falk, pp. 15 ff.

2.1.4 Confirmation Bias

Confirmation bias is understood to mean that people have a tendency to seek confirmation of a hypothesis they have already formed. In this process, it is not just the case that information confirming the hypothesis is more readily perceived and given greater weight; somewhat like a positive testing strategy, confirming data are actively sought while data that would falsify the hypothesis are screened out.[35] The danger posed by the confirmation bias for judicial decision making is self-evident: as soon as a judge is even remotely convinced of the negligence of a defendant, factors that might contradict this conclusion are unconsciously screened out of the judgement process. Therefore jurists should never commit themselves prematurely to a hypothesis, and certainly not publicly. Litigators must also be aware that their view of the case is often distorted by their biased perspective as a representative of one party's interests.[36]

2.1.5 Egocentric Bias

Two phenomena which correspond closely and can be summed up under the heading "egocentric biases" are those of unrealistically high hopes and unrealistic overassessment of one's own capabilities: "overoptimism" and "overconfidence":

> People routinely estimate, for example, that they are above average on a variety of desirable characteristics, including health, driving, professional skills, and the likelihood of having a successful marriage.[37]

The *overoptimism* bias means that people overestimate the probability of good things happening to them, and underestimate the risk that anything bad will happen to them. An example of this unrealism is the optimism of most car-drivers that they are less likely than others to be involved in a motoring accident – unrealistic because if that were the case, the statistical average probability would also have to be lower.[38] An explanation for this unrealistic optimism is often *overconfidence*. For instance, most drivers consider themselves to be more skilled than average, which gives them the feeling that in the worst case scenario – which, incidentally, tends to be screened out in any case – they could prevent an accident with a skilful driving manoeuvre.

Even judges are not immune to overconfidence. According to a study by Guthrie, Rachlinski and Wistrich, 87.7% of judges in the USA consider their own competence to be above average.[39] In Switzerland, according to a similar study by

[35] Schweizer, p. 178.

[36] Schweizer, p. 273.

[37] Guthrie, Rachlinski and Wistrich, pp. 811 f.

[38] Englerth, p. 95.

[39] Guthrie, Rachlinski and Wistrich, p. 814.

Schweizer, the figure is as high as 92%.[40] The overconfidence of judges is perhaps even a necessity for a functioning legal system. For the very knowledge of the possibility of cognitive biases such as anchoring or hindsight bias could actually make judges less secure about their own decision making. A certain overconfidence may counteract such insecurity. Apart from that, most people prefer a self-confident, decisive judge to one who is anxious and uncertain.[41]

2.2 Debiasing

The question that now arises is how one should react to cognitive biases, or how they can be prevented (known as "debiasing"). One problem is that these appear to be extremely *robust phenomena*. This is particularly true of hindsight bias, which – as we have seen – plays an important role in the assessment of the foreseeability of incidents of damage.

> Because the bias is so deeply ingrained into the human judgement process, psychologists have been unable to develop a way to induce people to make unbiased *ex post* judgements of *ex ante* probabilities.[42]

It is also known that raising awareness about the anchoring effect makes little or no difference.[43] Even if people recognize the invalidity of an arbitrary anchor, they will nevertheless use it as a starting point for their subsequent judgement. So even if the judge knows that the mere occurrence of an injury by no means directly implies that it was foreseeable or the fault of the injurer, there is nevertheless a great danger that his judgement will be biased in this direction. If *awareness-raising and warning* do not necessarily protect against biased judgements, the question is how else the problem can be addressed. Uppermost considerations in this regard are *institutional precautions* which counteract or compensate for cognitive biases.

Firstly, it would certainly be helpful for judges to establish *transparency* in relation to the basis of their decision-making and their deliberations. They have to explain transparently which circumstances of a sequence of events lead them to draw which conclusions. By this means, judgements will be more susceptible to scrutiny and constructive criticism, on the one hand. On the other hand, this procedure has the advantage that judges thoroughly analyse their own deliberations and, ideally, recognize where any bias might have crept into their judgement. It is questionable whether a ruling by a panel of judges offers effective protection against biased judgements. Studies have shown that group decisions are influenced as much

---

[40] Schweizer, pp. 268 f.

[41] Schweizer, p. 270.

[42] Rachlinski, p. 98; Kamin and Rachlinski, p. 92.

[43] Wilson, Houston, Etling and Breeke, p. 397.

by knowledge of the outcome of a course of events as decisions made by individuals.[44] All the same, a panel of judges on the bench and the associated discourse of the judges could promote the required transparency with regard to the basis for rulings.

Another approach consists of more "adoption of multiple perspectives" in order to be more accountable for the existence of judgement biases.[45] The court should always bear in mind that although the course of events in a certain case went one way and not another, it could equally well have taken a different course. When the outcome of the course of events is known, those circumstances that are indicative of it are more consciously perceived than circumstances that would have indicated an alternative outcome. This hindsight and confirmation bias can be reduced if the person reaching the judgement has to explain how alternative sequences of events that did not actually occur might equally have come to pass.[46] Consequently the judge must also seek active reasons militating against the occurrence of the event.[47]

For all the eagerness to alleviate the impacts of cognitive biases, a degree of caution is also advisable. If people believe they can recognize their own judgement biases, or recognize them correctly, there is in fact a danger that the attempt to rectify them will fail. Unnecessary corrections or *overcorrections* are the result:

> Even in the rare instances in which people believe that their judgments are biased, they may not successfully debias these judgments. In fact, their corrected judgments might be worse than their uncorrected ones.[48]

For instance, the attempt to envision alternative courses of events in order to correct the hindsight bias may only reinforce the impression that the version of events that occurred was inevitable.[49]

Another possible response to cognitive biases consists in appropriate *framing of positive law*.[50] In this way, systematic judgement biases in legislation, for instance, could be anticipated and their effect compensated *ex ante*. Various authorities take the view with regard to fault-based liability that hindsight bias could be avoided in the assessment of fault by making reference to rules formulated *ex ante*:

> A reliable judgement of reasonable care that was made before an accident has occurred is sometimes available. Many customary practices within an industry or a profession are attempts to identify precautions that constitute a reasonable level of care. So are many government safety regulations. In such cases, rather than conducting an openended inquiry into whether a defendant took reasonable precautions, a court could assess whether the defendant complied with the relevant custom or regulation. Under this rule, only a determination

[44] Stahlberg, Eller, Maass and Frey, p. 56.

[45] Guthrie, Rachlinski and Wistrich, p. 822.

[46] Chapman and Johnson, p. 134.

[47] Kamin and Rachlinski, p. 100.

[48] Wilson, Centerbar and Breeke, p. 191.

[49] Schwarz and Vaughn, p. 113.

[50] Guthrie, Rachlinski and Wistrich, pp. 828 f.

> that the defendant had not complied with the *ex ante* norm could support a conclusion that the defendant was negligent.[51]

Rachlinski's proposal is to be welcomed but needs refinement in various respects.

In the categories of causal liability, it is sufficient for particular standardized circumstances to exist, such as an objective duty of diligence and some particular disregard of this duty; causal liability is not dependent on fault. In the case of simple causal liability, the injurer has an option known as the exception proof – he can provide evidence that he complied with the prescribed standard of care or that the damage would have occurred even if due care had been taken, and escape liability.[52] Fault-based liability is a different matter. It is not directly related to having committed some breach of duty of a certain kind. The inner justification of liability for negligence lies rather in the accusation that the injurer failed to prevent some foreseeable and therefore also preventable damage.

Swiss adjudication fundamentally acknowledges both state and certain private standards as a yardstick for the degree of care to be taken in the individual case.[53] In the domain of fault-based liability, however, it is not possible to infer negligence solely from non-compliance with such behavioural standards; equally, compliance with such standards does not rule out any kind of negligence.[54] In fault-based liability, it is not the particular disregard of an objective duty of care that justifies the *responsibility* of the injurer, but the accusation of not having prevented a foreseeable instance of harm:

> The diligence that one must expend in order to avoid the reproach of negligence is not a special duty stipulated by law but the outflow of this general responsibility for the success that is foreseen to be at least possible.[55]

State or private standards of behaviour concretize the required level of diligence insofar as they contain a statement on the kind of behaviour that must reasonably be expected to result in the occurrence of some injury.[56] Judges can and should make reference to these when they must judge whether the occurrence of the injury was foreseeable for the injurer or not. The injurer escapes the reproach of negligence if he can demonstrate that the incident was not foreseen by him, and nor could it have been foreseen by the average reasonable person. Evidence of non-adherence to safety regulations, rules and custom and practice is therefore only an indication that

---

[51] Rachlinski, p. 105.

[52] Rey, N 880 ff.

[53] Schweizer, pp. 225 f.

[54] Rey, N 872 ff.

[55] "Die Sorgfalt, die man aufzuwenden hat, um den Vorwurf der Fahrlässigkeit zu vermeiden, ist keine besondere, durch die Rechtsordnung statuierte Pflicht, sondern der Ausfluss dieser generellen Verantwortung für den mindestens als möglich vorausgesehenen [. . .] Erfolg." Oftinger and Stark, § 5 N 56.

[56] Oftinger and Stark, § 5 N 58.

there must have been some expectation of an injury; it is indicative of negligence but does not imply it.[57]

One may ask how much of a problem cognitive biases really are. Biases in judgement are of course unproblematic with regard to *legal certainty.* Since the cognitive biases are systematic and everybody is basically subject to them, they pose no danger to the constancy and uniformity of the application of law. The discrepancy between postulated and realized law could – as mentioned – possibly be compensated by anticipation of judgement biases.

If efficiency is seen as a criterion in liability – which is of course disputed[58] – one may ask what consequences cognitive biases have on the structure of incentives and on efficiency. According to the "Hand rule", the efficient standard of care is defined as follows: if B represents the costs of the injurer's precautions, P the probability of damage and L the likely magnitude of damage, then an injury can be seen as having breached the duty of care provided that: $B < P * L$. Under the Hand rule, liability for negligence begins at precisely the point where the expected value of damage exceeds the cost of avoidance. If the cost of avoiding damage would amount to more than the expected value of the potential damage, on the other hand, then no such liability arises for failing to take the appropriate precautions. In more general terms, according to the Hand rule, a particular action is only required if it is efficient, i.e. if it generates more benefits than costs for society.[59] An unduly strict assessment of negligence would give rise to an excessive level of precautions (or an unduly low level of activity) in the potential injurer.

Finally it should be mentioned that in tort law – at least with regard to the efficiency aspect – one could also rely on *one bias compensating for another.* As mentioned, people tend towards the assessment that they are far less likely than other people to be affected by adverse events, which cannot in any way be realistic otherwise the average probability would also be lower. The lower precautions associated with overoptimism are compensated by the overestimation of probability by the court passing judgement *ex post.* Overoptimism and hindsight bias would thus be cancelled out in the ideal scenario.[60]

3 Empathy

Closely related if different is the question of whether a judge should possess empathy and whether this does not jeopardize impartiality and ultimately also the rationality of judgements. While many jurists are by no means so conscious of the influence of cognitive bias on the adjudication process, they see empathy as a spectre that has no place in judicial decision making.

---

[57] Brehm, N 185 on Art. 41, Swiss Code of Obligations.

[58] On the so-called bilateralism critique, see Mathis, pp. 100 f.

[59] For a more extensive discussion, see Mathis, p. 97 ff.

[60] Jolls, Sunstein and Thaler, pp. 38 f.

"Empathetic judging" as proclaimed by Barack Obama, as a presidential candidate on his electoral campaign, sparked vigorous debate in the American legal world. Obama said he wished to see people with heart and empathy on the bench of the United States Supreme Court, who were in a position to empathize with the socially disadvantaged.

> [W]e need somebody who's got the heart, the empathy, to recognize what it's like to be a young teenage mom [... and] to understand what it's like to be poor, or African-American, or gay, or disabled, or old.[61]

The question is naturally whether such a judge can still judge impartially. In order to answer this question, it is first necessary to establish what meaning attaches to the terms "impartiality" and "empathy" – quite generally, and in the specific context of judicial decision making.

According to Birnbacher, the *impartial standpoint* is characterized by the fact that it gives no weight to spatial, temporal, genealogical or social distance as a criterion for moral differentiation.[62] A judicial decision fulfils the requirement of impartiality on the one hand by complying with the procedural standards laid down in positive law and the choice and application of objectively justified material rules, and on the other hand through consideration of the relevant objective viewpoints of an applied situation.[63] Absolute objectivity cannot be demanded, however:

> The judge cannot blind out his own personality since he must, after all, make a decision. If the element of the personal decision could be eliminated from the determination of what is right, if what is right were objectively given and applicable in an absolute sense, then the demand for impartiality would be pointless.[64]

A judgement can be held up as impartial if it can be justified vis à vis all the parties affected.[65] The point at which it can no longer be justified is when it gives the interests of one party undue weight to the detriment of the other without being able to provide *objective reasons* for doing so. Empathy is no objective reason.

The word empathy goes back to the German concept "*Einfühlung*" and was used and developed primarily in the writings of Theodor Lipps.[66] An early translation from the German into English gave rise to the English term "empathy",[67] which in modern psychology denotes the capacity to put oneself in other people's position. It can also be referred to as "conceptual perspective taking". This refers to the ability

---

[61] Senator Barack Obama, quoted after Bandes, p. 135.

[62] Birnbacher, p. 414.

[63] Lohmann, p. 447.

[64] "Der Richter kann seine eigene Persönlichkeit nicht ausblenden, da er ja entscheiden muss. Könnte in der Bestimmung des Richtigen das Moment der persönlichen Entscheidung eliminiert werden, wäre also das Richtige in einem absoluten Sinne objektiv gegeben und zur Anwendung zu bringen, wäre die Forderung nach Unparteilichkeit sinnlos." Lohmann, p. 448.

[65] Lohmann, p. 454.

[66] See Theodor Lipps, *Grundlegung der Ästhetik*, and id., 'Einfühlung und ästhetischer Genuß'.

[67] Curtis, p. 11.

to perceive others as people with their own goals, interests and abilities.[68] An emotional and a cognitive aspect of empathy can be differentiated. *Emotional empathy* allows us to feel what another person is feeling. *Cognitive empathy* allows us to discern what another person is feeling.[69]

The ability to put oneself in other people's position is generally evaluated as a positive quality. This circumstance must not blind us to the fact that empathy can be utilized for very different purposes. An evolutionary benefit of empathy is thought to be that of understanding competitors in order to eliminate them more efficiently.[70] Someone with empathy can understand their counterpart more easily and therefore engage more effectively with their desires and interests. But it is also empathy that helps the sadist to find out how to torture his victim with particular cruelty.[71]

Now the question is what role empathy can and should play in judicial proceedings. Must judges not simply apply the law neutrally without empathizing with the parties affected? The opinion of Jessica Weisner points in this direction:

> Lady Justice doesn't have empathy for anyone. She rules strictly based upon the law and that's really the only way that our system can function properly under the Constitution.[72]

Richard A. Posner however considers such a position legalistic and describes empathy in "How Judges Think" (2008) as an important quality of the judge:

> [A] major factor in judicial decisions [. . .] is 'good judgement', an elusive faculty best understood as a compound of empathy, modesty, maturity, a sense of proportion, balance, a recognition of human limitations, sanity, prudence, a sense of reality, and common sense.[73]

Lynne N. Henderson similarly advocates incorporating empathy into judicial decisions, claiming that the opponents of empathy were basing their position on a stunted understanding of rationality:

> The avoidance of emotion, affect, and experimental understanding reflects an impoverished view of reason and understanding – one that focuses on cognition in its most reductionist sense. This impoverished view stems from a belief that reason and emotion are separate, that reason can and must restrain emotion, that law-as-reason can and must order, rationalize, and control.[74]

Susan A. Bandes in her essay "Empathetic Judging and the Rule of Law" (2009) dealt thoroughly with Obama's comments and the corresponding responses. She contends that empathy cannot simply be equated with fellow feeling. Empathy is far more a *cognitive competence* than a feeling:

> Empathy is a capacity, not an emotion. It differs from sympathy or compassion, both of which are emotions. Empathy entails understanding another person's perspective. Sympathy

---

[68] Henderson, p. 1581.

[69] Ekman, p. 249.

[70] Breithaupt, p. 175.

[71] Williams, p. 91.

[72] Jessica Weisner, quoted after Bandes, p. 138.

[73] Posner, p. 117 (author's emphasis).

[74] Henderson, p. 1576.

> is a feeling for or with the object of the emotion. Empathy does not require, or necessarily lead to, sympathy. Empathy, unlike sympathy, does not necessarily lead to action on behalf of its object, or the desire to take action on his behalf.[75]

Bandes brings cognitive empathy to the forefront but, wrongly, dismisses the emotional aspect of empathy. She finds it misleading to discuss whether judges should have empathy or not, considering it a matter of fact that their own feelings would enter into their reaction to the feelings of the parties. For her, the more critical point is that judges should *be accountable* for their own capacity for empathy, its limits and any "blind spots".[76] Henderson argues along similar lines. She points to the phenomenon that we tend to develop more empathy for people from our own social group.

> The reality of empathy is that we are more likely to empathize with people similar to ourselves, and that such empathic understanding may be so automatic that it goes unnoticed: elites will empathize with the experience of elites, men empathize with men, women with women, whites with whites. I would call this 'unreflective' empathy.[77]

Once again, this is a context in which "biases" come into play. Martin L. Hoffmann mentions the "familiarity bias" and the "here-and-now bias". On the one hand, we have more empathy with people with whom we are more familiar; on the other hand, empathy is especially strong for people who are spatially and temporally closest to hand. The "familiarity bias" can be broken down further into "in-group bias", "friendship bias" and "similarity bias": we have more empathy for group members (e.g. people of the same ethnicity), for friends and for people with certain shared characteristics (e.g. members of the same sex).[78]

Empathy can therefore indeed represent a danger to the impartiality of judicial judgments, particularly when the judge does not reflect critically on his or her own feelings. Even more dangerous, however, is the denial of empathy, since in this way its risks are masked and judgements are based on an individual, narrow-minded point of view which one mistakenly believes to be generally accepted.[79] A critically-reflected ability to empathize, on the other hand, can be of great benefit to law, as Henderson rightly contends:

> [E]mpathy is a form of understanding, a phenomenon that encompasses affect as well as cognition in determining meanings, it is a rich source of knowledge and approaches to legal problems – which are, ultimately, human problems.[80]

Conscious analysis of the intentions and interests of the parties and one's own prejudices is, on the one hand, an indispensable prerequisite for a judgement that

---

[75] Bandes, p. 136.

[76] Bandes, p. 135.

[77] Henderson, p. 1584.

[78] Hoffmann, pp. 206 ff.

[79] McLane Wardlaw, pp. 119 f.

[80] Henderson, p. 1576.

fulfils its intended function, namely to bring about reconciliation between the competing interests of the parties to the proceedings.[81] On the other hand, the empathetic judge is also accountable for the effects of this process on his own emotions. Empathy in the application of law does not therefore mean that judges allow private intentions, interests and prejudices to influence their judgement directly or even to take the place of judgement.[82] On the contrary, they should enable judges to develop and preserve the necessary conscious distance between the aims of the parties and the judges' own feelings.

Markus Müller refers in this connection to the role of empathy in the communications of officialdom with ordinary citizens. In his view, the authorities should show understanding for the personal situation of citizens and communicate with them empathetically, i.e. with intuitive understanding, which presupposes a capacity for comprehension and appropriate verbal expression.[83] Müller sees the recognizable effort to try and understand the other person as a strong sign of esteem, which is undermined by the use of formulaic language, schematic structures and stereotypes; these are likely to give members of the public the impression that the authority has not taken their needs and wishes seriously, or even heard them at all. This, he warns, is detrimental to acceptance and also leaves the bitter impression that officialdom is incapable of understanding.[84] Only if public communication between the state and citizens, but also between public authorities, is based on understanding, empathy and respect are the basic requirements met for the legitimation of state action.[85]

4 Conclusion

As our analysis has established, cognitive biases pose a serious threat to the rationality of the judicial adjudication process, and the avoidance of such errors appears to be a difficult matter. Specifically, courts are obviously susceptible to the anchoring effect and to hindsight errors even when they are aware of the phenomenon of cognitive biases. Awareness-raising and warnings are therefore relatively ineffective means of addressing cognitive biases. A more advisable approach is certainly for the courts to establish transparency concerning the basis for their decisions and deliberations, thereby opening up their judgements to constructive criticism.

Having a number of judges on the bench seems to be of little help, since collective decisions are as likely as individual decisions to be subject to cognitive biases. All the same, it however could promote the required transparency with regard to the basis for rulings. A further recommendation is to give more consideration to alternative perspectives, a practice which can be particularly helpful in avoiding

[81] O'Grady, pp. 9 ff.

[82] McLane Wardlaw, pp. 117 f.

[83] Müller, pp. 131 f.

[84] Müller, p. 132.

[85] Müller, p. 151.

hindsight errors and confirmation bias. Finally, another form of "debiasing" is possible through the detailed framing of positive law. By taking the relevant judgement bias into account in legislation, the error could be anticipated and compensated . In tort law, for instance, an unduly strict assessment of negligence influenced by hindsight bias and anchoring could be counteracted by the precise *ex ante* formulation of standards of due diligence.

One may ask how much of a problem cognitive biases really represent. Since judgement biases are systematic, neither the constancy and coherence of the application of law nor legal security are ultimately jeopardized. The problem is that a discrepancy can arise between postulated and implemented law. Moreover, from the viewpoint of efficiency, inefficient incentive structures can be created – as the example of tort law has shown. Perhaps in this case, however, one might rely on the fact that different cognitive biases compensate for one another. For example, the unrealistic optimism of potential injurers may cancel out the hindsight bias of the courts passing judgement on the damages case.

The situation is a little different with regard to the role of empathy in judicial decision making. Empathy can also cause distortions, although these are located more on the affective than on the cognitive plane. Thus we have more empathy towards people with whom we are more familiar, or who are closer to us in space and time. If empathy is understood not only as an emotion but also as a cognitive competence to recognize what others are feeling, and if one is also in a position to reflect critically on one's own feelings, then empathy in the application of law might be quite beneficial. The main purpose of the application of law is to resolve conflicts between people. This is easier to accomplish if judges are in a position to put themselves in other people's positions and to critically question their own personal standpoints. Professionally applied empathy can therefore only increase the rationality and, ultimately, the legitimacy of judicial judgements and state actions.

Bibliography

Bandes, Susan A., 'Empathetic Judging and the Rule of Law', in *Cardozo Law Review De Novo*, Vol. 133 (2009), pp. 133 ff.

Birnbacher, Dieter, *Analytische Einführung in die Ethik* (2nd edn., Berlin and New York, 2007)

Brehm, Roland 'Die Entstehung durch unerlaubte Handlungen, Art. 41-61 OR', in Heinz Hausheer and Hans Peter Walter (eds.), *Berner Kommentar, Kommentar zum schweizerischen Privatrecht, Band VI: Das Obligationenrecht* (3rd edn., Bern, 2006), 1. Abteilung: Allgemeine Bestimmungen, 3. Teilband, 1. Unterteilband [Part 1: General Provisions, Volume 3, Subvolume 1]

Breithaupt, Fritz, *Kulturen der Empathie* (Frankfurt a.M., 2009)

Chapman, Gretchen B. and Johnson, Eric J. 'Incorporating the Irrelevant: Anchors in Judgments of Belief and Value', in Thomas Gilovich, Dale Griffin and Daniel Kahneman (eds.), *Heuristics and Biases. The Psychology of Intuitive Judgment* (New York, 2002), pp. 120 ff.

Christensen-Szalanksi, Jay J. and Fobian-Willham, Cynthia, 'The Hindsight Bias: A Meta-Analysis', in *Organizational Behavior and Human Desicion Processes*, Vol. 48 (1991), pp. 147 ff.

Curtis, Robin, 'Einführung in die Einfühlung', in Robin Curtis and Gertrud Koch (eds.), *Einfühlung. Zur Geschichte und Gegenwart eines ästhetischen Konzepts* (Munich, 2009), pp. 11 ff.

Englerth, Markus, 'Behavioral Law and Economics – eine kritische Einführung', in Christoph Engel, Markus Englerth, Jörn Lüdemannand and Indra Spiecker genannt Döhmann (eds.), *Recht und Verhalten* (Tübingen, 2007), pp. 60 ff.

Ekman, Paul, *Gefühle lesen. Wie Sie Emotionen erkennen und richtig interpretieren* (2nd edn., Heidelberg, 2010)

Falk, Ulrich, 'Urteilsverzerrungen: hindsight bias und anchoring, Einleitende Fragen zu einem interdisziplinären Problem', in Brodersen, Kai (ed.), *Antike Kultur und Geschichte, Vol. 11: Vincere Scis, Victoria Uti Nescis, Aspekte der Rückschauverzerrung in der Alten Geschichte* (Berlin, 2008), pp. 9 ff.

Fischhoff, Baruch, 'Hindsight ≠ Foresight: The Effect of Outcome Knowledge on Judgement Under Uncertainty', in *Journal of Experimental Psychology: Human Perception and Performance*, Vol. 1 (1975), pp. 288 ff.

Guthrie, Chris, Rachlinski, Jeffrey J. and Wistrich, Andrew J., 'Inside the Judicial Mind', in *Cornell Law Review*, Vol. 86 (2001), pp. 777 ff.

Hastie, Reid, Schkade, David A. and Payne, John W., 'Juror Judgements in Civil Cases: Hindsight Effects on Judgements of Liability for Punitive Damages', in *Law and Human Behavior*, Vol. 23 (1999), pp. 597 ff.

Henderson, Lynne N., 'Legality and Empathy', in *Michigan Law Review*, Vol. 85 (1987), pp. 1574 ff.

Hoffmann, Martin L., *Empathy and Moral Development. Implications for Caring and Justice* (Cambridge, 2000)

Jolls, Christine, Sunstein, Cass R. and Thaler, Richard, 'A Behavioral Approach to Law and Economics', in Cass R. Sunstein (ed.), *Behavioral Law and Economics* (Cambridge, 2000), pp. 13 ff.

Jolls, Christine, 'Behavioral Law and Economics', in Alain Marciano (ed.), *Law and Economics* (New York, 2009)

Jungermann, Helmut, Pfister, Hans-Rüdiger and Fischer, Katrin, *Die Psychologie der Entscheidung. Eine Einführung* (3rd edn., Heidelberg, 2010)

Kahneman, Daniel and Tversky, Amos, 'Prospect Theory: An Analysis of Decision under Risk', in *Econometrica*, Vol. 47 (1979), pp. 263 ff.

Kamin, Kim A. and Rachlinski, Jeffrey J., 'Ex Post ≠ Ex Ante, Determining Liability in Hindsight', in *Law and Human Behavior*, Vol. 19 (1995), pp. 89 ff.

LaBine, Susan J. and LaBine, Gary, 'Determinations of Negligence and the Hindsight Bias', in *Law and Human Behavior*, Vol. 20 (1996), pp. 501 ff.

Lipps, Theodor, *Grundlegung der Ästhetik. Erster Teil* (Hamburg, 1903)

Lipps, Theodor, 'Einfühlung und ästhetischer Genuß', in *Die Zukunft*, Vol. 54 (1906), pp. 100 ff.

Lohmann, Georg, 'Unparteilichkeit der Moral', in Lutz Wingert and Klaus Günther (eds.), *Die Öffentlichkeit der Vernunft und die Vernunft der Öffentlichkeit. Festschrift für Jürgen Habermas* (Frankfurt a.M., 2001), pp. 434 ff.

Mathis, Klaus, *Effizienz statt Gerechtigkeit? Auf der Suche nach den philosophischen Grundlagen der Ökonomischen Analyse des Rechts* (3rd edn. Berlin, 2009), [English edition: Mathis, Klaus, *Efficiency Instead of Justice? Searching for the Philosophical Foundations of the Economic Analysis of Law* (New York, 2009)]

McLane Wardlaw, Kim, 'Umpires, Empathy, and Activism: Lessons from Judge Cardozo', in *Notre Dame Law Review*, Vol. 85 (2010), pp. 1629 ff.

Müller Markus, *Psychologie im öffentlichen Verfahren. Eine Annäherung* (Bern, 2010)

Oftinger, Karl and Stark, Emil, *Schweizerisches Haftpflichtrecht. Vol. I: Allgemeiner Teil* (5th edn., Zurich, 1995)

O'Grady, Catherine, 'Empathy and Perspective in Judging: The Honorable William C. Canby, Jr.', in *Arizona State Law Journal*, Vol. 33 (2001), pp. 4 ff.

Posner, Richard A., *How Judges Think* (Cambridge MA and London, 2008)
Rachlinski, Jeffrey J., 'A Positive Psychological Theory of Judging in Hindsight', in Cass R. Sunstein (ed.), *Behavioral Law and Economics* (Cambridge, 2000), pp. 95 ff.
Rey, Heinz, *Ausservertragliches Haftpflichtrecht* (4th edn. Zurich, Basel and Geneva, 2008)
Schwarz, Norbert and Vaughn, Leigh Ann, 'The Availability Heuristic Revisited: Ease of Recall and Content of Recall as Distinct Sources of Information', in Dale Griffin Thomas Gilovich and Daniel Kahneman (eds.), *Heuristics and Biases, The Psychology of Intuitive Judgment* (New York, 2002), pp. 103 ff.
Schweizer, Mark, *Kognitive Täuschungen vor Gericht. Eine empirische Studie* (Zurich, 2005)
Simon, Herbert A., 'Rational Decision Making in Business Organizations', in *American Economic Review*, Vol. 69 (1979), pp. 493 ff.
Stahlberg, Dagmar, Eller, Frank, Maass, Anne and Frey, Dieter, 'We Knew It All Along: Hindsight Bias in Groups', in *Organizational Behavior and Human Decision Processes*, Vol. 63 (1995), pp. 46 ff.
Tversky, Amos and Kahneman, Daniel, 'Judgment under Uncertainty: Heuristics and Biases', in *Science*, New Series, Vol. 185 (1974), pp. 1124 ff.
Williams, Bernard, *Ethics and the Limits of Philosophy* (Cambridge MA, 1985)
Wilson, Timothy D., Houston, Christopher E., Etling, Kathryn M. and Breeke, Nancy, 'A New Look at Anchoring Effects: Basic Anchoring and Its Antecedents', in *Journal of Experimental Psychology: General*, Vol. 125 (1996), pp. 387 ff.
Wilson, Timothy D., Centerbar, David B. and Breeke, Nancy (2002), 'Mental Contamination and the Debiasing Problem', in Thomas Gilovich, Dale Griffin and Daniel Kahneman (eds.), *Heuristics and Biases. The Psychology of Intuitive Judgment* (Cambridge, 2002), pp. 185 ff.

Part II
Law and Sustainability

Our Responsibility Towards Future Generations

Paolo Becchi

The future, for a long time, was a concept to which philosophers and jurists paid little attention. Barely any of the great thinkers of the past classed it as relevant to human action. Even Kant, who shows some signs of taking the problem of generational succession into consideration, bases his ethics on the presumption of the individual acting in the *hic et nunc*, for whom the future clearly remains a remote concern. The constant advancement of technology is now turning this conception upside down. The first to recognize the practical implications of this circumstance, and to suggest that the technological civilization called for a new ethics, was Hans Jonas. The following essay therefore begins with an appraisal of Hans Jonas, before proceeding to make a – *prima facie* – surprising comparison with Herbert Hart. Examination of the deficiencies of both perspectives opens up the avenue of transcendence which leads to a new orientation, at least an outline of which is put forward in conclusion.

1 Introduction

The future of humankind was long regarded as a consideration of no great relevance in human lives. Equally, the upshot of modernity's radical break with the past was ultimately to focus its interest firmly upon the present. The future was left to chance, to fate, or perhaps to divine providence. So alien to juristic thinking was the idea of caring about future generations that Savigny even doubted it possible to draft a perfect law code "for a succeeding and less fortunate age", commenting that "an age is seldom disposed to be so provident for posterity".[1]

In philosophy, Kant the "philosopher of history" shows the earliest signs of taking the problem of generational succession into consideration. He starts out from

[1] Cf. von Savigny, pp. 42 f.

P. Becchi (✉)
University of Lucerne, CH-6002 Lucerne, Switzerland; University of Genoa, Liguria, Italy
e-mail: paolo.becchi@unilu.ch; paolo.becchi@unige.it

K. Mathis (ed.), *Efficiency, Sustainability, and Justice to Future Generations*, Law and Philosophy Library 98, DOI 10.1007/978-94-007-1869-2_4,

a teleological view of nature (which may seem old-fashioned today, perhaps, but – as I will show – will nevertheless be crucial to my line of argument), whereby all natural capacities must evolve completely to their natural end. This means, in relation to the human being (the only rational being on Earth, our given purpose being the use of reason as an end), that humankind's abilities never attain complete evolution within the single individual, but in the generational succession within the human race: for each individual human being would need an infinitely long lifespan in order to learn to make full use of all his abilities. "Since Nature has set only a short period for his life, she needs a perhaps unreckonable series of generations, each of which passes its own enlightenment to its successor in order finally to bring the seeds of enlightenment to that degree of development in our race which is completely suitable to Nature's purpose."[2] Although life expectancy nowadays is much longer than in Kant's times, this in no way weakens the power of his words. Thus, despite these considerations, even for Kant the future of humankind is rooted in the "realization of Nature's secret plan"[3] rather than in responsible human action.

The meaning of history, the advancement from generation to generation, is not the result of a human plan although we undoubtedly play our part in it. Kant therefore seems surprised that earlier generations would instinctively strive for something from which only later generations will benefit. The idea that emerges here might be characterized as that of solidarity between the generations, which does indeed have its puzzling aspects.[4] We are seemingly presented with a historical teleology in which the hope of progress depends not on the concrete action of individuals but on "a wisdom from above (which bears the name of Providence if it is invisible to us)".[5] In this way, the future is perpetually steered in the direction of betterment.

Thanks to this way of thinking, the existence of the human race in the world is regarded as a fact for which no further proof is required. Each of us dies, but the race as a whole is immortal. Although this same dimension of finality occurs in the formulation of the categorical imperative which refers to the idea of humanity as an "end in itself", this imperative is fundamentally directed to the individual; more precisely, the individual acting in the *hic et nunc*, for whom the future remains a remote concern. The horizon of the individual's actions does not extend beyond the foreseeable course of his own life.

---

[2]Quoted as in Kant, *Universal History*, p. 13.

[3]Ibidem, p. 21.

[4]Ibidem, p. 14: "It remains strange that the earlier generations appear to carry through their toilsome labor only for the sake of the later, to prepare for them a foundation on which the later generations could erect the higher edifice which was Nature's goal, and yet that only the latest of the generations should have the good fortune to inhabit the building on which a long line of their ancestors had (unintentionally) labored without being permitted to partake of the fortune they had prepared. However puzzling this may be, it is necessary if one assumes that a species of animals should have reason, and, as a class of rational beings each of whom dies while the species is immortal, should develop their capacities to perfection."

[5]Cf. Kant, *Conflict*, p. 169.

But this conception is turned upside down in the present epoch by the constant advancement of technology, which extends its dominion over all things in being and reduces existence to nothingness. The age of technology is the age of the will to power and of nihilism, and – as we know – prompted Martin Heidegger (from as early as the end of the 1930s, under the influence of Ernst and Friedrich Georg Jünger,[6] and then in the course of the 1950s) to raise questions about the destructive power of technology and the danger it presents not only to the environment but also to the human race. Yet Heidegger only commented on the theoretical dimension of this danger.[7] The coupling of "physics and responsibility" that he mentions in *On the Way to Language* essentially remains a hollow platitude.[8]

The first thinker to recognize the nature and practical implications of technology, and to emphasize the necessity for a new ethics for the technological civilization, was Hans Jonas in his foundational work *Das Prinzip Verantwortung (The Imperative of Responsibility)* published in 1979 (1984). For this reason, I would like to begin with Jonas and then proceed to consider Herbert Hart and his *Concept of Law*. At first glance, readers may be surprised by the juxtaposition of these two names. Indeed, the one (originally a continental European and even a proponent of a "rehabilitation of metaphysics") seems to have nothing to do with the other (an uncompromising analyst). However, I will try to show how analysis of the question of future generations not only brings together very different kinds of authors, but also has implications for both ethical and juristic reflection.[9] In my conclusion, the question I want to pose is whether it is possible to solve the problem of future generations on the ethical and juristic levels alone, or whether one should not in some way or another reintroduce the category of the sacred, on which the predominant technico-scientific approach has seemingly inflicted irreparable damage.

2 An Ethics of Responsibility for the Future Generation: The Paradigm of Hans Jonas

The qualitatively new nature of our action has also opened up a new spatiotemporal dimension of ethics. The repercussions of our actions are no longer confined to one space and one time but are becoming extremely powerful and indeterminate: through the actions we take here and now, we can not only influence the lives of

[6]Cf. F. G. Jünger, p. 232 (own trans.): "Man no longer masters the mechanical law which he himself has set in train; this law masters him". Certainly the greater influence on Heidegger was the work of Ernst Jünger. See E. Jünger, *Mobilmachung* and *Arbeiter*.

[7]Cf. Heidegger, 'Technik'. For commentary, cf. Schirmacher; Seubold. Among recent publications: Luckner.

[8]Cf. Heidegger, *Language*, pp. 102 f.: "'Physics and responsibility' – that is a good thing, and important in today's crisis. But it remains double-entry accounting behind which there is concealed a breach that can be cured neither by science nor by morality, if indeed it is curable at all."

[9]Of the recent literature, reference should at least be made to the comprehensive essay collection published by Bifulco and D'Aloia, to Birnbacher and Schicha, and to Schlothfeldt.

other people elsewhere, but even the future of generations yet to come, those not yet in existence. The most important change in the traditional framework comes from the vulnerability of both nature and humankind. For with the extension of technological power, the relationship between humankind and nature has changed to such a degree that the environment and even the human race itself are endangered by human manipulations in the field of genetic technology. This sober realization gives rise to Jonas' proposal.[10]

At first sight, however, his proposal seems ambiguous: on the one hand, it is meant as a response to a particular factual situation, so that it represents an ethic of responsibility *for* the technological civilization; on the other hand, the way in which it introduces the category of responsibility signals the intention to construct the underpinnings for generally binding moral principles. On closer scrutiny, though, both points of view are seen to be interlinked, since Jonas' real intention is something different: namely, an ontological foundation of ethics, which *alone*, in the final instance, is in a position to justify the survival of the human race and hence of future generations. These three connected aspects will nevertheless be discussed separately, if also briefly, here.

Jonas' reference to responsibility makes us think first, inevitably, of Weber's distinction between the ethics of responsibility and the ethics of conviction.[11] These are two approaches which are characterized by two different but perhaps not necessarily incompatible guiding principles. The ethics of conviction is absolute and unconditional, and does not concern itself with the consequences that a particular action may have: the action must be good in and of itself, irrespective of the attainment of any desired end. Kant's categorical imperative seems to exemplify such an "ethic of duty for duty's sake". The ethic of responsibility, in contrast, evaluates an action on the basis of the foreseeable consequences that it can bring in its wake and for which responsibility must be borne.

It is obvious that this second approach must necessarily consider the relationship between means and ends: an action is closely connected to the end it aims to accomplish, and the acting individual must be able to foresee the possible consequences of his or her action. For Jonas, the means are the totality of resources at the disposal of the present generation, and the ends are the survival of the human race on Earth and the perpetuation of life in general. Hence he proposes an ethics for the technological civilization, i.e. an ethics that has arisen out of a particular historical situation – our own – and which claims validity for precisely this situation.

From the very first pages of his work, it is nonetheless apparent that Jonas understands responsibility in a different sense. Despite all the critiques that he addresses to Kant, his responsibility inevitably calls to mind the categorical imperative, i.e. a

[10]One of the first writings in which Jonas anticipates the main thesis of the book for which he became renowned is 'Technology', pp. 35 ff. (later in the anthology Jonas, *Philosophical Essays*, pp. 3 ff.). For an extensive discussion of *Imperative*, Müller is still useful. For a historical contextualization cf. Gethmann-Siefert, pp. 171 ff. Jonas' significance has not escaped one of the first juristic monographs on the topic: Saladin and Zenger, pp. 26 ff.

[11]Cf. Weber. For a relevant discussion (in relation to Jonas) cf. Nepi, pp. 95 ff.

general principle that is valid for any time and any place. Jonas offers several formulations of this principle. Here we can concentrate on his first formulation: "Act so, that the effects of your action are compatible with the permanence of genuine human life."[12] Jonas is undeniably introducing a consequentialist line of argument, although here he appears to make reference to the second formulation of the Kantian categorical imperative: "Act so that you treat humanity, whether in your own person or in that of another, always as an end and never as a means only".[13] It is rather astonishing that Jonas does not compare his principle of responsibility with this Kantian formulation of the categorical imperative, and pauses to pass criticism of the first formula only.

Whether or not we accept his criticism of Kant,[14] it has to be asked whether Jonas' reformulation does not itself run the risk – to express it in Kantian terms – of transforming the categorical into a hypothetical principle. For Jonas' formulation implicitly contains the following train of thought: "If you desire the permanence of genuine human life on Earth, then you must incorporate the future integrity of human life on Earth into your decisions." But in this way, the model of the new imperative would be purely hypothetical and not categorical. Where Kant essentially sees such "pragmatic imperatives" as targeted at happiness and welfare, for Jonas the objective is much more limited: it is survival. If all that Jonas had offered were merely a new justification of this hypothetical imperative, indeed he would not have gone much beyond Weber's approach, other than to adapt it to the new implications of human action. In that event, however, Jonas would not have developed a new ethics but just a series of prudent instructions with which we should comply as we go about realizing our own aims.

Without a doubt, many of his instructions regarding various bioethical themes that he examined[15] can be explained in this way, i.e. as prudent and precautionary measures, which may be able to guarantee human existence and life on Earth in a particular situation – our own – in which both are at risk. To read Jonas in this sense would be limiting, however, because it would disregard the fact that he prefixes that imperative, which could actually go no further than a simple justification of goal-directed, strategic action, with another imperative which makes express reference to a metaphysical view of existence, thereby guiding thoughts from ethics towards ontology.

---

[12]Jonas, *Imperative*, p. 11.

[13]Quoted as in Kant, *Foundations*, p. 47.

[14]Jonas' criticism of Kant seems to me to go no further, in essence, than the critisicm levelled at Kant's ethics by Hegel, who referred to it as "empty formalism". On this, see: Hegel, 'Glauben', p. 89; 'Jacobi', pp. 24 ff.; 'Grundlinien', Vol. VII, §§ 133–137, pp. 250 ff.; 'Enzyklopädie', Vol. VIII, §§ 53–54 (and supplementary notes), pp. 138 f.; 'Vorlesungen', Vol. XX, pp. 366 ff.; secondary literature: Knox, pp. 70 ff.; Ritter, pp. 281 ff.; Cesa, pp. 149 ff.; Ameriks, pp. 179 ff.

[15]Cf. Jonas, 'Technik'. On the complex relations between theory and practice in the *Imperative*, I refer to my essay 'Diskrepanz', pp. 239 ff.

The first imperative is namely "*that* there be a mankind".[16] "The Kantian distinction between the hypothetical and categorical imperative, intended for the ethic of contemporaneity, applies as well to the ethic of responsibility for the future which we are here trying to define. The hypothetical (of which there are many) says: *if* there are human beings in the future – which depends on our procreation – *then* such and such duties are to be observed by us towards them in advance; but the categorical commands only *that* there be *human beings*, with the accent equally on the *that* and on the *what* of obligatory existence. For me, I admit, this imperative is the only one which really fits the Kantian sense of the categorical, that is, the unconditional. Since *its* principle is not as with Kant, the self-consistency of reason giving to itself laws of conduct, that is not an idea of *action* [. . .], but is rather the idea of possible agents in general, for whom it claims that such *ought to exist* (fallible as they must be), and is thus 'ontological', that is, an idea of *being* – it follows that the first principle of an 'ethic of futurity' does not itself lie *within* ethics as a doctrine of action [. . .], but within *metaphysics* as a doctrine of being, of which the idea of Man is a part."[17]

This first "ontological" imperative permits the justification of responsibility as the principle of our actions. Thus, both Weber's and Kant's schemata are ultimately inappropriate, since the unconditional duty to act responsibly has an extra-moral, which is to say an ontological, basis.

Why should each of us act in such a way as to favour the perpetuation of the human race on Earth? And above all, why should this duty apply to all, unconditionally, as an absolutely binding ought? The answer is not to be found in ethics but in ontology. But what kind of ontology is being referred to? Jonas attempts to set out an ontological justification for his categorical imperative by drawing on Aristotelian metaphysics. Aristotelian teleology is the original wellspring of his endeavour.[18] Being has a destined purpose: since being has created life, being has shown itself to have at least one aim, which is life itself. And life must not be confused with Descartes' *res extensa*, since even in its simplest forms it shows a preference for its own preservation and self-affirmation.

In the striving towards an end, Jonas sees "a fundamental self-affirmation of being [. . .], which posits it *absolutely* as the better over against nonbeing".[19] And the first end that being strives towards is that of self-preservation. This superiority of purpose over lack of purpose, of being as against nonbeing, forms the fundamental ontological principle that enables Jonas to interpret the inherent purpose of being not just as a fact but also as an intrinsic value. If being is to be favoured over nonbeing, it means that the end towards which being strives, i.e. its own preservation, is also a value to be protected, and as such represents a duty for humankind: the duty of all duties, the ought of being. This establishes a seamless transition from the

[16] Jonas, *Imperative*, p. 43.

[17] Jonas, *Imperative*, pp. 43 f.

[18] As the most challenging of Jonas' works proves very well. Cf. *Phenomenon*.

[19] Jonas, *Imperative*, p. 81.

level of ontological reflection to the deontological level. There is no discontinuity, because axiology becomes part of ontology. Being is better than nonbeing, having ends is better than having none, and the end that being strives towards is that of perpetuation: this end is life itself, which manifests itself in plants, in animals and in its most perfect form in human beings. And the existence of such a life is something good per se, since the existence of sentient life is better than its absence, the existence of conscious life is preferable to the lack of it, and life itself is better than non-life. Now, if it is good that there is such a thing as this life, then it is also our duty to ensure its perpetuation.

Thus, Jonas derives an "ought" from the "is" of being, but this is not to say that he thereby commits a naturalistic fallacy. He does not interpret being as it is considered in the modern natural and social sciences as a value-neutral fact. Taking up the Aristotelian teleology once again, Jonas understands being itself as an end in and of itself with an intrinsic value. The "ought" derived from the "is" of being is one which, in reality, is already present in being, since being, as opposed to nonbeing, itself possesses a value. Jonas also provides a concrete example in which the difference between existence (is) and obligation (ought) would be eliminated: the existence of the newborn, "whose mere breathing uncontradictably addresses an ought to the world around, namely, to take care of him."[20] Here, the naked "is" immediately coincides with the "ought". The most delicate and defenceless of all beings, the newborn, the future generation that has become present, demands that responsible action be taken to care for it, becomes the archetype of that form of obligation which is intuitively based on the objective value of being.

Many objections can be levelled at this attempt to base ethics on an onto-teleological, metaphysical premise, and indeed, many have been.[21] And Jonas himself has conceded to some extent that his opponents are right, when a few years after the publication of *The Imperative of Responsibility*, he admitted in an essay that his "argument [could] do no more than rationally justify an *option*".[22] To stay with the example of the newborn, it might be possible to ask, for instance, how its parents should behave in the event that it suffered from such extreme deformities that it would certainly die, and perhaps soon. Should they fight for its existence to the bitter end, or – at least in this case – would its nonbeing be preferable to its being? One might reply that Jonas' train of thought does not concern *that* particular newborn individual, but *the* newborn as such. The being and the nonbeing do not relate to that particular child but, ultimately, to the being and the nonbeing of future

---

[20]Jonas, *Imperative*, p. 131. And Jonas continues: "I mean strictly, just this: that here the plain being of a *de facto* existent immanently and evidently contains an ought for others, and would do so even if nature would not succour this ought with powerful instincts or assume its job alone."

[21]Starting from the objection argued by Apel, which is certainly one of the most thorough: 'Verantwortung', pp. 179 ff. And from among the recent publications, cf. Hirsch Hadorn, pp. 218 ff.

[22]Cf. Jonas, 'Grundlegung', p. 140 (own trans.): "Yet ultimately my argument can do no more than rationally justify an *option* which it presents with its inherent persuasive force to the thinking individual. Unfortunately I have nothing better to offer. Perhaps a future metaphysics will be able to."

generations. Indeed, on the basis of his train of thought we can arrive at a justification of our duty to ensure the perpetuation of the human race on Earth. The example nevertheless demonstrates the limits of Jonas' approach very well: on the foundation of that ontological premise, he can only guarantee the survival of the human race, but not the equal right to such survival for every human individual.

This – almost biological – imperative of conservation is not sufficient to justify an ethic nor an ethics for the technological civilization. This is because, from a simple biological point of view, the perpetuation of the human race, considering the present situation of overpopulation and resource scarcity, could also be ensured by sacrificing a large proportion of the population for the sake of the future generations. Of course, Jonas would certainly not be in agreement with this cynical perspective, but on the basis of his ontological principle that humankind should exist, it cannot be entirely ruled out.

Furthermore the (fair) demand to guarantee the perpetuation of the human race cannot skirt the issue of considering the conditions in which individuals will have to live in future. The perpetuation of human life on the planet would be pointless if it came at the price of human degradation.

In summary, Jonas' attempt to derive a new version of the categorical imperative from the first, almost biological imperative ("*that* there be a mankind") founders: he can only ensure the biological perpetuation of the race but not – as Jonas very much wanted to – the permanence of genuine human life on Earth.[23]

3 Survival as the Objective of Law: A New Interpretation of Herbert Hart's "Minimum Content of Natural Law"

As emphasized in the foregoing, Jonas attempted to make the perpetuation of the human race on Earth the pivotal theme in his justification of a new ethics for the technological civilization. This theme – surprisingly enough – already had a certain relevance in one of the most important works of legal philosophy of the last century: *The Concept of Law* by Herbert Hart.[24] In this work as elsewhere Hart advocated

[23] In a radio interview, Jonas finds himself forced to admit this. His interviewer (Eike Gebhardt) had asked him: "Do you consider pure survival to be a value in itself, the greatest good so to speak, quite irrespective of *what* survives? Since you only argue *ex negativo*: might we call your position an 'ethics of avoidance'?" And Jonas replied: "I would accept your expression 'ethics of avoidance'; not as the definitive formula for an ethics, but in the sense of a temporary priority [...]. I do not say that there can or ought to be a definitive ethics; but the situation has made it imperative that the avoidance of extreme dangers and extreme evil be prioritized, while the search for the best, the ideal or the exemplary human life, both for the individual and for society, steps into the background for the time being." (Own trans., cf. 'Naturwissenschaft', pp. 208 ff.).

[24] Cf. Hart, *Concept*, pp. 255 ff. In German-speaking legal philosophy, the argument of the minimum content of natural law has barely been discussed. (Cf. the brief presentation in the monograph by Eckmann, pp. 48 ff.). Among older publications, mention is made of Passerin d'Entrèves, pp. 312 ff., and Cattaneo, pp. 673 ff. Of the more recent publications, see: Rivaya, pp. 65 ff.; Orrego, pp. 287 ff. and Epstein, pp. 219 ff. In Italy two additional papers have been published on

legal positivism, understood as a theory which disputes the existence of a necessary conceptual connection between law and morality. But in Chapter IX of this work, Hart also refers to the "minimum content of Natural Law", putting forward an argument first signalled in his earlier essay, *Positivism and the Separation of Law and Morals,*[25] in which the author clearly takes a different stance from the version of legal positivism propounded by Kelsen.

For Kelsen, the law, seen in functionalistic terms, is actually nothing more than a specific social technique (its use being essentially to ensure compliance with certain behavioural norms within a society with the threat of coercive measures), and the standard for assessing its effectiveness is reduced to reviewing the coherence found in the technique itself. Consequently, Kelsen can derive from it the effectual formula that "[a]ny content whatever can be law".[26]

Since Hart, in contrast, considers the survival of the human race as the first objective of law, he again appropriates the teleological view of nature also assumed by Jonas, and with astonishing similarity, steers the attention towards survival as the "proper end of human activity".[27] Strictly speaking, both Jonas' philosophical reflection and Hart's juridical reflection are based on the essential *telos* of survival which living beings generally strive towards.

When Hart views human survival as the primary objective of law, he is nevertheless attributing particular content to the legal system; namely, that its aim is to ensure the survival of humankind on Earth. In this perspective, Hart gives a new interpretation of modern natural law, drawing on precisely those authors, such as Hobbes and Hume who – albeit with differences – saw "in the modest aim of survival the central indisputable element which gives empirical good sense to the terminology of Natural Law."[28]

Although Hart holds fast to the thesis of legal positivism, according to which a norm legally enacted within a legal system is law even when it is unjust, he asserts (or at least, the inference can be drawn) that any legal system which does not presuppose a few elementary truths is not worthy of the name. For in that case, the function of preserving human survival, which represents the primary objective of law, would not be fulfilled. Even before one ascertains "*how* men should live together, we must

the theme: Ricciardi, pp. 221 ff. and Maestri, pp. 240 ff. Finally, mention is made of a work with a strong bearing on our problematique, Bifulco, pp. 61 ff.

[25]Cf. Hart, 'Positivism', pp. 593 ff.

[26]Cf. Kelsen, p. 56. Kelsen continues: "there is no human behaviour that would be excluded simply by virtue of its substance from becoming the content of a legal norm."

[27]See Hart, *Concept*, p. 187. Based on a few purely biological facts like the primary human needs (food, rest, etc.) Hart draws the self-evident conclusion: "it is the tacit assumption that the proper end of human activity is survival, and this rests on the simple contingent fact that most men most of the time wish to continue in existence. The actions which we speak of as those which are naturally good to do, are those which are required for survival; [. . .]."

[28]Hart, *Concept*, p. 187. Here, Hart sets Hobbes and Hume against followers of the Aristotelian tradition because the latter did not confine themselves to survival as such but saw survival "as merely the lowest stratum in a much more complex and far more debatable concept of the human end or good for man".

assume that their aim, generally speaking, is to live."[29] There are some elementary truths – Hart calls them truisms – which are inherent to human beings as such and are associated with their natural environment and the ends they strive towards. From these truisms, some universally binding behavioural principles can be derived, which Hart simply defines as "ranked as natural laws discoverable by reason".[30] In this regard, there are some truisms which are inherent to human nature as such. Their content amounts to the necessary core that must always be at the heart of law, if it is to possess at least the minimum objective of survival that people pursue in order to live together.

The first four truisms impose on individuals duties which they cannot change, so that the resulting norms are static. Let us consider them more closely: 1) "Human vulnerability". Based on their naturally endowed capacities, people are normally vulnerable. The most typical commandment of law and morality, "You shall not kill", can be explained precisely through that mutual vulnerability. Were this not the case – if, for example, humans had impenetrable armour, then this commandment would not be needed. 2) "Approximate equality". Although people have different physical strength, mobility and intellectual capacities at their disposal, they are not subdivided into superior and inferior beings but recognize each other as equals and consequently see the necessity of a system of compromises and reciprocal non-violence. The Hobbesian roots[31] of these evident truths are immediately discernible: in the state of nature, "every man has a right to everything; even to one another's body". As long as this state persists, nobody can be certain (however strong and clever he may be) of being able to live for the entire span of time which nature normally grants human beings; hence the necessity of securing peace by establishing a state.

Hart's other two static truths are both already to be found directly in Hume's *Treatise of Human Nature.*[32] These are 3) "limited altruism", which seems to hark back to the "limited generosity" which Hume had thought natural to man, and to which could be added the "asocial sociability" *(ungesellige Geselligkeit)* mentioned by Kant. Human nature is neither demonic nor angelic.[33] Humans are driven neither by the desire to annihilate each other nor by selfless, reciprocal love. Thus, human altruism exists but it is limited and inconstant: furthermore, aggressive tendencies can always gain the upper hand if they are not controlled, so that a system of mutual

---

[29]Hart, *Concept*, p. 188.

[30]Hart, *Concept*, p. 188. See also p. 189: "In considering the simple truisms which we set forth here, and their connexion with law and morals, it is important to observe that in each case the facts mentioned afford a *reason* why, given survival as an aim, law and morals should include a specific content. The general form of the argument is simply that without such a content laws and morals could not forward the minimum purpose of survival which men have in associating with each other."

[31]Cf. Hobbes, p. 66.

[32]Cf. Hume, p. 315.

[33]One critique of Hart in this point is delivered by Barry, but it does not strike at Hart's central point, cf. Barry, p. 158.

forbearance is necessary and also possible. 4) "Limited means" coincides with what Hume had called the "slender means, which she [Nature] affords to the relieving these necessities",[34] i.e. the basic needs that humans have for food, clothing, shelter, and all those things not found in unlimited supply in nature. This fact makes some form or other of property as well as norms on its protection necessary.

To these four truths Hart added a fifth, which is dynamic for it enables individuals to create new duties and to modify their impacts. Hart defines this in terms which are not intuitively comprehensible: "limited understanding, and limited strength of will". Most people are intelligent enough to understand that it is necessary to comply with the rules arising from the first four truths. But some are not. Therefore sanctions are needed to ensure that those who would obey voluntarily are not unjustly harmed by the behaviour of free-riders. For there are always some who would take advantage of the social system without subjecting themselves to the duties arising therefrom. Here, once again, we return to Hume[35] and particularly to his three fundamental rules which are indispensable for the existence of every society: the stability of possession, the transference of property by consent, and the performance of promises. Human vulnerability, approximate equality, limited altruism, limited means, limited understanding and limited strength of will are qualities of humans and of their life-worlds which make some rules seem necessary, and without which the purpose of survival could not be fulfilled.

All this is associated with survival, according to Hart, since the scarcity of resources results in norms which enable the accumulation of the necessary resources for survival. Giving, exchanging and selling the products of work creates a system of new obligations, but since altruism is not unlimited and since we cannot expect everyone to understand the collective interest of keeping promises, sanctions are necessary to reinforce general adherence to the norms that ensure peaceful coexistence.

Hart claims that these five truisms are necessary in order to understand any social organization: they amount to the minimum content shared by law and morality, without which no form of human coexistence would be possible. Therefore, at the end of his exposition, he comes to a critique of Kelsen's positivist thesis that "law may have any content."[36] Regardless of whether or not his thesis remains compatible with legal positivism, given that Hart is normally numbered among the legal positivists, the centrality of the concept of survival in this author's work must be stressed. One might counter that observation with the argument that Hart is essentially analysing a function of law which has always been inherent to it: ensuring peaceful coexistence between humans, and consequently individual survival. After all, Jhering, too, spoke of *Law as a Means to an End*:[37] to his mind, however, the "end" was that of securing the conditions for the permanence of society, whereas Hart is concerned

[34]Cf. Hume, p. 189.

[35]Cf. Hume, pp. 210 ff.

[36]Hart, *Concept*, p. 195.

[37]von Jhering.

with ensuring the survival of human beings as a natural desire of any given individual. Be that as it may, why should we attach such significance to this chapter of Hart's book when the matter of "responsibility towards future generations" in which we are interested does not appear in his list of five elementary truths?

It comes as no surprise that it does not, because at that time – *The Concept of Law* was published in the early 1960s – approaches to the "future generations" problem had yet to emerge in the philosophical debate. Let it be emphasized, nevertheless, that on the very basis of that "modest teleologism" of which Hart speaks, Jonas (knowing nothing of his famous forerunner) constructed his "new" ethic of responsibility over the course of the following decade. Certainly, Hart is inspired by a different aim; perhaps by the intention of overcoming groundless dichotomies, such as the one between Natural Law and legal positivism, which often make the discussion about the properties of law unclear. But the way in which he tackles the subject opens up new pathways in the debate that has preoccupied philosophers so greatly and for so long. Hart's ideas about those truisms concerning human nature and the world in which humans live show that, as long as these are valid, there are certain behavioural rules which every social organization must adhere to if it wants to stay alive. Thus, as long as human nature and the world in which humans live retain their natural form, the elementary truths that Hart talks about retain their validity. In the chapter mentioned he insists on this point, both at the beginning of his exposition and at the end, where he declares that the truth of the five elements is contingent on human beings and the world they live in retaining the salient characteristics which they have now.[38]

At this point, all Hart wishes to show is an empirical version of Natural Law based on Hobbes and Hume. But the way in which he treats the theme, i.e. the emphasis on preserving human nature and the persistence of a natural environment that safeguards humankind's survival, also creates an implied avenue onto the theme of our responsibility towards future generations.

If human survival is accepted as the objective of law, this presupposes the concept of humankind as it has been and as it is now, but not as it could become. In other words, the survival we are striving for is the survival of the human race in the form it presently takes. And the truths we talked about concern humans, their natural environment and the ends they pursue. For Hart, then, the crux is the preservation of those conditions without which the survival of humans on Earth could not be guaranteed.

Now supposing that ensuring the survival of the human race is the primary task of law, then the law's scope – given present-day technology, which endangers such survival both directly (through genetic manipulation) and indirectly (through environmental degradation) – is broadened to encompass future generations. These take on greater importance in that the interest of people today in preserving the conditions in which they have always lived gains a future dimension. The task of preserving and safeguarding human life in the given biological state, in a natural

[38]Hart, *Concept*, p. 195.

environment that permits its perpetuation, compels us also to take into consideration the survival of the human race in the future, and grant it the right to the integrity of its genetic stock and to a natural environment that is sufficiently healthy to support its survival. Nothing more than this (and even this is no small thing to ask nowadays), especially since the aims pursued by humans are very diverse.

From a legal-philosophical point of view, one might perhaps see in this the preconditions for a new "minimal" Natural Law, which no longer opposes legal positivism but complements it.[39] Hartian positivism presupposes the existence of a few very general principles which are not metaphysical in themselves since they are the result of empirical consideration of human nature. And precisely on this empirical basis it would be possible to give our responsibility towards future generations a purely rational foundation.

It is not my intention to discuss Hart's undoubtedly original approach in every detail. One critical aspect, however, merits at least some mention. It concerns the logical status of those elementary truths on which the thesis of the minimum content of natural law is founded. Actually, Hart barely comments on this: according to him, they are neither "definitions", nor "ordinary statements of fact", but "a third category of statements", namely those whose truth happens to be contingent upon whether humans and the world in which they live go on retaining the salient characteristics that they possess now.[40]

Hart always had a sense of the versatility of language and of the different functions it can perform. In one of his early writings, he showed, for example, how language is not reduced only to the descriptive and the prescriptive function, but also has an ascriptive function[41] which consists of the attribution of responsibility and rights. Within the context in which we are interested, however, it remains unclear which function he ascribes to those statements on which the minimum content of natural law is founded. In saying that those elementary truths are not definitions, he seems to exclude them from the definition of law; in saying that they are not statements of ordinary facts, he also seems keen to rule out that they must necessarily be present in existing juridical organizations. Thus, Hart has only told us what they are *not*, and is thus exposed to at least two possible objections. On the one hand, we do not understand the way in which the above statements differ from statements of fact; on the other hand – in the event that they should contain a value judgement – they might infringe Hume's law forbidding the derivation of normative statements from factual premises.

Regarding the first point, the factual statements in Hart seem to refer exclusively to the present and to the past. Statements which concern the survival of the human race in the future – insofar as they imply no value-judgement – are limited to describing the fact that humans normally want to go on living. The fact that, under certain conditions the human race does not die out, is not sufficient to differentiate

[39]In this regard, cf. Bifulco, pp. 61 ff.

[40]Cf. Hart, *Concept*, p. 195.

[41]Cf. Hart, 'Ascription', pp. 145 ff.

such statements from the statements of fact; just as on the level of logic, one cannot differentiate the statement "it is raining" from the statement "it will rain". Both are factual statements. The first is true if it actually rains; the second could be untrue if the weather forecast were wrong; but all this permits no logical distinction between the two types of statements.

It would be a different matter if the preservation of the human race were to imply a value-judgement. Admittedly, unlike Jonas, Hart does not expressly assert that the aim of living (and continuing to live) is a value, but his thoughts on the minimum content of Natural Law would corroborate, or at least not preclude, this interpretation as well. Although there are people who under certain circumstances would rather die than live, in most cases precisely the opposite is true. One can therefore conclude that the aim of survival is something good. Were this the case, however, Hart would be committing the logical error of deriving a value judgement (that the preservation and protection of human life is something good) from a factual premise (that there are humans). But this is a liberty that Hart – who, unlike Jonas, emphasizes the fundamental distinction between is and ought – cannot allow himself to take.[42]

Either the survival of the human race is something good that should be protected (in which case, however, Hart would be introducing a value-judgement into the description and defending a view at odds with his meta-ethical, non-cognitivist approach), or these elementary truths convey something very banal, which is that as far as we know, humans have shown a preference for associating to protect their own lives rather than to form suicide pacts. Hart is too insistent that the aim of survival implies a contingent necessity to be able to accept the first hypothesis. If only the second hypothesis remains valid, one might wonder whether it suffices to justify our responsibility towards future generations. That is doubtful. On this foundation, it is not possible to justify the idea that there should be people like us in future, but only the idea that up to now, people have preferred living over dying, and if they would like to continue living in future, they must consider the consequences of their actions on future generations. If their aim as humans is to continue living, then those attributes which characterize human nature and the conditions which make this possible should also be guaranteed in future.

According to Hart's theory, it is not ultimately possible to formulate any absolute duty towards future generations. Yet in the final analysis, the closing balance is not entirely negative. The elementary truths that he expounds provide good grounds on which to arrive at a few specific findings. If we want to realize at least that minimal *telos* towards which humans naturally strive, i.e. to carry on existing, then present generations should act in such a way as not to modify the conditions which allow humans to continue living as they do now (such as a healthy environment and the integrity of genetic stock). But equally, some might not wish to pursue this end, and as a consequence, the nihilistic will to create a post-human reality would greatly

[42]Whether Hart is impermissibly deriving an "is" from an "ought" is a matter of controversy in the literature; see references in n. 24 above.

endanger the future of humanity. Hart's truisms would be of little use to counteract this, in my opinion.

Finally, there is a danger both in Hart and in Jonas that forms of coexistence based on fundamental inequality have to be tolerated in order to salvage at least the minimum requirement of survival. Although survival is a necessary condition for human development, in its own terms it is compatible with the possibility of many people living in degrading conditions. In order to save future generations, one might endanger at least a share of those alive now.

4 Future of the Species and the Avenue of Transcendence: Tentative Outlines

Perhaps we must do as Jonas does and critically explore the problem further, in even greater breadth than he does. Instead of responsibility towards (past, present or future) generations, we should do better to speak about responsibility towards the image of humankind that has been handed down to us since time immemorial. And we can also continue to defend it in future – against current "creative" attempts to manipulate it – since every such vision corresponds to a timeless concept of the human being. We can only think about a responsibility towards human beings who will be alive 1000 years from now on the condition that these human beings will be like us in nature. But can law and morality explain the rigidity, the immutability of this vision of humankind, or does this not create room for something that transcends them both?

One of the great ideal-types on which modern Western culture is founded was described by Max Weber in exemplary fashion as a process of rationalization and demystification of the world. This modern model of secular self-understanding has not only meant the dissolution of metaphysics into the individual sciences but has also confined religion and – more generally – moral values and norms to the private sphere of individual conscience. The counterpart to "scientistic positivism", oriented according to the rationality paradigm of value-neutral science, is the loss of the public dimension of religion which, in a similar way to ethics, has been reduced to a private matter. In contrast to scientific-technical rationality, ethical and religious choices are individual decisions which stem from personal, ultimately irrational feelings.

Ethics has long endeavoured to depart from this schema. Mention need only be made here of John Rawls' theory of justice, Hans Jonas' principle of responsibility, and Karl-Otto Apel's discourse ethics. These authors represent the pinnacle of efforts to establish a rational ultimate justification for ethics. Their attempts at a "rehabilitation of practical philosophy" (Jonas only partly included) took place against a background which brooked no transcendental presuppositions. To all intents and purposes, it seemed that the Good Lord had exhausted his function and no further doubts should be cast on the Weberian paradigm, at least as far as religion was concerned. Ethics could take to the public arena without further ado, but religion was consigned purely to the private sphere.

The undeniable breakthrough of religiosity into the public sphere in various forms in recent years – as correctly foreseen by Habermas – has precipitated a crisis in this mode of thinking. What emerges from this new phenomenon might be called "rehabilitation of political theology". For many, this means a risk-laden retreat to the past, and even a major peril for democracy. In my opinion, however, democracy nowadays is imperilled by different matters entirely. For instance, when a word from an American rating agency is all that it takes to force the European Union to its knees! Be that as it may, hardly a day goes by without the press printing some plea or another for secular rationality, in which ideological bric-a-brac that is of no use for an understanding of reality is revamped in neo-enlightenment manner. The crucial question remains: is the West seriously endangered by this rebirth of political theology or is the paradigm of secularization, taken to the extreme, now teetering on the brink of collapse?

It is claimed that the absence of God, or at least his remoteness, far from human events, will be filled nihilistically by transferring his (lost) omnipotence onto *homo creator.* The human will becomes the double of the divine will. The search for freedom from any external dependency which the modern era has doggedly pursued is unmasked today in its extreme radicalism: a delusion of omnipotence which has dammed humankind to endure a situation of radical ontological loneliness in the universe.

The dream of human self-conquest, and of liberation from a divine transcendence that is increasingly perceived as a competitor, has mutated into the nightmare of an absolute freedom which creates the monster of a will to power, which first sets its face against external nature and then against human nature itself. The paradoxical consequence of this absolutization of immanence is the abasement of the human being. As Nietzsche put it, "man seems to have got himself on an inclined plane – now he is slipping faster and faster away from the centre".[43] Man has gone from governing subject to governed object, a passive and defenceless means for the realization of ever more sophisticated and devastating technological experiments.

This is the project of genetic engineering and its many flatterers, and at the same time the greatest peril of our age, which truly calls into question the survival of human beings on Earth. We are all connected, like a web, but also caught in this web. Being everywhere and nowhere, we have already lost our sense of place. Now we are also in the process of losing our sense of time.

The human race seems to have reached the terminal stage of its evolution. Already, the pump is primed for a new reality: the post-human, the creation of a new race by direct interference with the genetic code of the existing race. Can anything be done to counteract this process of self-destruction? An ethics of responsibility founded on the future of humankind can certainly provide an orientation

[43]Quoted as in Nietzsche, p. 155.

point, but Jonas himself remarks[44] that in times of extreme danger something different is needed, something that goes even beyond metaphysics. Perhaps the avenue of transcendence – repressed and yet always present – could evince a meaningful, motivating force once again. For can the inviolability of the human being be justified in any other way than through a rediscovery, possibly in the form of a negative theology, of the category of the sacred which was divested too hastily in the name of enlightenment?

Before the human being was made the subject, by Descartes, he had never found his true extent within himself, in the *fundamentum inconcussum* of his own self-assurance, but in the religious sphere. In order to prevent today's process of absolutization of the human being, this myth of the "superman", from culminating paradoxically in his total destruction, one should rediscover the religious sense of one's own limits and of awe before the sacred as the ultimate horizon of meaning. Rationality alone is not enough; it must be nourished by something that it cannot produce itself.

Of course, it would be naive to believe that one could combat the growing spread of nihilism with a new synthesis of faith and knowledge. In the first centuries of our culture, it was indeed the synthesis realized by theology between Christianity and Platonism which conquered the first wave of nihilism (that of Gnosticism). Today, however, after the radical skepticism of Nietzsche and Heidegger, this appears impossible. We must come to terms with the absence of God: continue to feel the presence of this absence. This is far from saying that "everything is possible", that evolution is now at the mercy of the "will to power" of a human being who would even gamble with his own future in order to take the place of God. It is rather the case that we cannot avoid living in a horizon of radical doubtfulness, in which there are no ultimate metaphysical certainties and guarantees, but only an unending search for the meaning of things. This meaning is not instilled in things by human beings; it is there, and the human being is the only being that can discover it.

God has not bequeathed us his creator role, but has created the human being "in his likeness" and thus bestowed on us that transcendental *dignitas* which allows us to take up a special position within nature. Before and beyond every value, perhaps pointing out the specificity of the human condition and of its ontological meaning will allow us to rein in or even halt the current insane course of the biotechnological society towards self-destruction. Perhaps this also represents an anchor-point of hope for future generations: for we have absolutely no right to rob them of the *dignitas*[45] that characterizes us, or to turn the human race into a relic in evolutionary history.

---

[44]Jonas, *Imperative*, p. 23. "It is moot whether, without restoring the category of the sacred, the category most thoroughly destroyed by the scientific enlightenment, we can have an ethics able to cope with the extreme powers which we possess today and constantly increase and are almost compelled to wield."

[45]Cf. Becchi, 'Puzzle', pp. 157 ff., and now in greater depth, Becchi, *Principio*.

Bibliography

Ameriks, Karl, 'The Hegelian Critique of Kantian Morality', in Bernard den Ouden (ed.), *New Essays on Kant* (New York, 1987), pp. 179 ff.

Apel, Karl-Otto, 'Verantwortung heute – nur noch Prinzip der Bewahrung und Selbstbeschränkung oder immer noch der Befreiung und Verwirklichung von Humanität?', in Karl-Otto Apel, *Diskurs und Verantwortung. Das Problem des Übergangs zur postkonventionellen Moral* (Frankfurt a.M., 1990; cited as: 'Verantwortung'), pp. 179 ff.

Barry, Brian, *Theories of Justice* (Berkeley, 1989)

Becchi, Paolo, 'Gibt es eine Diskrepanz zwischen Theorie und Praxis des Prinzips Verantwortung?', in Ralf Seidel and Meiken Endruweit (eds.), *Prinzip Zukunft: Im Dialog mit Hans Jonas* (Paderborn, 2007), pp. 239 ff. (cited as: 'Diskrepanz')

Becchi, Paolo, 'Das Puzzle der Menschenwürde', in Paolo Becchi, Christoph B. Graber and Michele Luminati (eds.), *Interdisziplinäre Wege in der juristischen Grundlagenforschung* (Zurich, Basel and Geneva, 2007), pp. 157 ff. (cited as: 'Puzzle')

Becchi, Paolo, *Il principio dignità umana* (Brescia, 2009; cited as: *Principio*)

Bifulco, Raffaele and D'Aloia, Antonio, *Un diritto per il futuro* (Naples, 2008)

Bifulco, Raffaele, *Diritto e generazioni future. Problemi giuridici della responsabilità intergenerazionale* (Milano, 2008)

Birnbacher, Dieter and Schicha, Christian, 'Vorsorge statt Nachhaltigkeit – ethische Grundlagen der Zukunftsverantwortung', in Dieter Birnbacher and Gerd Brudermüller (eds.), *Zukunftsverantwortung und Generationensolidarität*, Schriften des Instituts für angewandte Ethik e.V., Vol. 3 (Würzburg, 2001), pp. 17 ff.

Cattaneo, Mario A., 'Il diritto naturale nel pensiero di H.L.A. Hart', in *Rivista internazionale di filosofia del diritto,* Vol. XLII (1965), pp. 673 ff.

Cesa, Claudio, 'Tra Moralität e Sittlichkeit. Sul confronto di Hegel con la filosofia pratica di Kant', in Valerio Verra (ed.), *Hegel interprete di Kant* (Naples, 1981), pp. 149 ff.

Eckmann, Horst, *Rechtspositivismus und sprachanalytische Philosophie. Der Begriff des Rechts in der Rechtstheorie H. L. A. Harts* (Berlin, 1969)

Epstein, Richard A., 'The not so minimum content of natural law', in *Oxford Journal of legal studies*, Vol. 2 (2005), pp. 219 ff.

Gethmann-Siefert, Annemarie, 'Ethos und metaphysische Erbe. Zu den Grundlagen von Hans Jonas' Ethik der Verantwortung', in Herbert Schädelbach and Geert Keil (eds.), *Philosophie der Gegenwart. Gegenwart der Philosophie* (Hamburg, 1993), pp. 171 ff.

Hart, Herbert L.A., 'The Ascription of Responsibility and Rights', in Antony Flew (ed.), *Essay on Logic and Language* (Oxford, 1951), pp. 145 ff. (cited as: 'Ascription')

Hart, Herbert L.A., 'Positivism and the Separation of Law and Morals', in *Harvard Law Review*, Vol. 71 (1958), pp. 593 ff. (cited as: 'Positivism')

Hart, Herbert L.A., *The Concept of Law* (London, 1961; cited as: *Concept*)

Hegel, Georg W.F., 'Grundlinien der Philosophie des Rechts oder Naturrecht und Staatswissenschaft im Grundrisse', in Eva Moldenhauer and Kurt Markus Michel (eds.), *Werke in zwanzig Bänden,* Vol. VII (Frankfurt a.M., 1969–1971), pp. 250 ff. (cited as: 'Grundlinien')

Hegel, Georg W.F., 'Enzyklopädie der philosophischen Wissenschaften im Grundrisse (1830)', in *Werke in zwanzig Bänden*, Vol. VIII (Frankfurt a.M., 1969–1971), pp. 138 f. (cited as: 'Enzyklopädie')

Hegel, Georg W.F., 'Vorlesungen über die Geschichte der Philosophie III', in *Werke in zwanzig Bänden*, Vol. XX (Frankfurt a.M., 1969–1971), pp. 366 ff. (cited as: 'Vorlesungen')

Hegel, Georg W.F., 'Glauben und Wissen, oder die Reflexionsphilosophie der Subjektivität in der Vollständigkeit ihrer Formen, als Kantische, Jacobische, Fichtesche Philosophie', in Hans Brockard and Hartmut Buchner (eds.), *Jenaer Kritische Schriften (III)* (Hamburg, 1986), p. 89 (cited as: 'Glauben')

Hegel, Georg W.F., 'Jacobi-Rezension', in id., *Berliner Schriften (1818–1831)*, ed. Walter Jaeschke (Hamburg, 1997), pp. 8 ff. (cited as: 'Jacobi')

Heidegger, Martin, 'Die Frage nach der Technik' (1950), in id., *Vorträge und Aufsätze* (Pfullingen, 1962), pp. 37 ff.; published in English as Martin Heidegger, *The question concerning technology, and other essays*, trans. and with an introduction by William Lovitt (New York, 1977; cited as: 'Technik')

Heidegger, Martin, *On the Way to Language* (= *Unterwegs zur Sprache*, 1959), trans. Peter D. Hertz (New York, 1971; cited as: *Language*)

Hirsch Hadorn, Gertrude, 'Verantwortungsbegriff und Kategorischer Imperativ der Zukunftsethik', in *Zeitschrift für philosophische Forschung*, Vol. 54 (2000), pp. 218 ff.

Hobbes, Thomas, *Leviathan* (1651; 5th edn., London, 1885)

Hume, David, *A Treatise of Human Nature* (1739/40), Vol. II, Passions – Morals (London, 1817)

Jonas, Hans, *The Phenomenon of Life. Towards a Philosophical Biology* (New York, 1966; cited as: *Phenomenon*)

Jonas, Hans, 'Technology and Responsibility. Reflections on the New Tasks of Ethics', in *Social Research*, Vol. 40 (1973), pp. 35 ff. (cited as: 'Technology')

Jonas, Hans, *Philosophical Essays. From Ancient Creed to Technological Man* (Chicago, 1974; cited as: *Philosophical Essays*)

Jonas, Hans, *The Imperative of Responsibility: In Search of an Ethics for the Technological Age* (Chicago and London, 1984, paperback edition 1985; cited as: *Imperative*)

Jonas, Hans, *Technik, Medizin und Ethik. Praxis des Prinzips Verantwortung* (Frankfurt a.M., 1985; cited as: *Technik*)

Jonas, Hans, 'Zur ontologischen Grundlegung einer Zukunftsethik' (1985), now in id. (ed.), *Philosophische Untersuchungen und metaphysische Vermutungen* (Frankfurt a.M. and Leipzig, 1992), pp. 128 ff. (cited as: 'Grundlegung')

Jonas, Hans, 'Naturwissenschaft versus Natur-Verantwortung', in Dietrich Böhler (ed.), *Ethik für die Zukunft. Im Diskurs mit Hans Jonas* (Munich, 1994), pp. 197 ff. (cited as: 'Naturwissenschaft')

Jünger, Ernst, *Die totale Mobilmachung* (Berlin, 1931; cited as: *Mobilmachung*)

Jünger, Ernst, *Der Arbeiter. Herrschaft und Gestalt* (Hamburg, 1932; cited as: *Arbeiter*)

Jünger, Friedrich G., *Die Perfektion der Technik* (1939) (2nd edn., Frankfurt a.M., 1949)

Kant, Immanuel, *Foundations of the Metaphysics of Morals* (= Grundlegung zur Metaphysik der Sitten, 1785), trans. Lewis W. Beck (Indianapolis, 1959; cited as: *Foundations*)

Kant, Immanuel, *Idea for a Universal History from a Cosmopolitan Point of View* (= Idee zu einer allgemeinen Geschichte in weltbürgerlicher Absicht, 1784), trans. Lewis W. Beck (Indianapolis, 1963; cited as: *Universal History*)

Kant, Immanuel, *The Conflict of the Faculties* (= Der Streit der Fakultäten, 1798) trans. and introduction by Mary J. Gregor, paperback edition (New York, 1979; cited as: *Conflict*)

Kelsen, Hans, *Reine Rechtslehre* (Vienna, 1960), quoted here from Hans Kelsen, *Introduction to the pure theory of law: a translation of the first edition of the Reine Rechtslehre or Pure theory of law*, trans. Bonnie Litschewski Paulson and Stanley L. Paulson (Oxford, 1997)

Knox, Thomas M., 'Hegel's Attitude to Kant's Ethics', in *Kant-Studien*, Vol. 49 (1957–58), pp. 70 ff.

Luckner, Andreas, *Heidegger und das Denken der Technik* (Bielefeld, 2008)

Maestri, Enrico, *Giudizi di esistenza. Deliberare sulla vita umana nella riflessione bioetica contemporanea* (Napoli, 2009)

Müller, Wolfgang E., *Der Begriff der Verantwortung bei Hans Jonas* (Frankfurt a.M., 1988)

Nepi, Paolo, *La responsabilità ontologica. L'uomo e il mondo nell'etica di Hans Jonas* (Rome, 2008)

Nietzsche, Friedrich, *On the Genealogy of Morals* (1887), ed. and trans. Walter A. Kaufmann (New York, 1989)

Orrego, Cristóbal, 'H. L. A. Hart's Understanding of Classical Natural Law Theory', in *Oxford Journal of Legal Studies*, Vol. 2 (2004), pp. 287 ff.

Passerin d'Entrèves, Alessandro, '"Un noyau de bon sens". À propos de la théorie du droit naturel chez H. Hart', in *Revue internationale de philosophie,* Vol. 17 (1963), n. 65, fasc. 3

Ricciardi, Mario, *Diritto e natura. H. L. A. Hart e la filosofia di Oxford* (Pisa, 2008)

Ritter, Joachim, 'Moralität und Sittlichkeit. Zu Hegels Auseinandersetzung mit der kantischen Ethik', in *Metaphysik und Politik. Studien zu Aristoteles und Hegel* (Frankfurt a.M., 1969), pp. 281 ff.

Rivaya, Benjamin, 'On theories of the minimum content of natural law – constructions or reconstruction', in *Rechtstheorie*, Vol. 32 (2001), pp. 65 ff.

Saladin, Peter and Zenger, Christoph A., *Rechte künftiger Generationen* (Basel and Frankfurt a.M., 1988)

Schirmacher, Wolfgang, *Technik und Gelassenheit. Zeitkritik nach Heidegger* (Freiburg im Breisgau and Munich, 1983)

Schlothfeldt, Stephan, 'Wer trägt die Verantwortung für zukünftige Generationen', in Carmen Kaminsky and Oliver Hallich (eds.), *Verantwortung für die Zukunft: zum 60. Geburtstag von Dieter Birnbacher*, Ethik in der Praxis, Vol. 22 (Berlin, 2006), pp. 37 ff.

Seubold, Günter, *Heideggers Analyse der neuzeitlichen Technik* (Freiburg im Breisgau and Munich, 1986)

von Jhering, Rudolf, *Der Zweck im Recht* (1877–1883), Volumes 1 and 2. Vol. 1 translated into English as *Law as a Means to an End*, trans. Isaac Husik (Boston, 1913)

von Savigny, Friedrich C., *Of the Vocation of our Age for Legislation and Jurisprudence* (= Vom Beruf unserer Zeit für Gesetzgebung und Rechtswissenschaft, 1814), trans. Abraham Hayward (2nd edn., London, 1831, reprinted 2002 by The Lawbook Exchange Ltd.)

Weber, Max, *The Vocation Lectures. Science as a Vocation, Politics as a Vocation* (= Wissenschaft als Beruf, 1917/1919. Politik als Beruf, 1919), ed. and with an introduction by David Owen and Tracy B. Strong, trans. Rodney Livingstone (Indianapolis, 2004)

Future Generations in John Rawls' Theory of Justice

Klaus Mathis

One question addressed by John Rawls in "A Theory of Justice" (1971) is that of justice between the generations. The question presents Rawls with certain difficulties which stem from the fact that in his theory, Hume's conditions of justice are spliced together with Kant's principle of universalization. The question of future generations strains this construct to breaking point. But even if this problem can be solved, Rawls' justification approach remains unsatisfactory, since he discusses intergenerational justice only under the aspect of just saving. It will be shown that this approach implies a concept of weak sustainability which is not a sufficient condition for the long-term preservation of vital life-sustaining resources.

1 Introduction

The subject of this essay is how John Rawls deals with the question of justice between generations in his writings. It begins with a discussion of the problems that intergenerational justice poses for contract theory (1). There follows a systematic account and critical appraisals of the various solutions proposed by Rawls, which culminate in the development of an alternative solution (2). In a further step, the argument is tied in with the sustainability debate and a distinction drawn between the concepts of strong and weak sustainability (3). The final section summarizes the conclusions of the inquiry (4).

As published in *Archives for Philosophy of Law and Social Philosophy (ARSP)*, Vol. 95 (2009), pp. 49 ff. With thanks to Franz Steiner publishing house for their kind permission to reprint this article here.

K. Mathis (✉)
University of Lucerne, Faculty of Law, Frohburgstrasse 3, P.O. Box 4466, CH-6002 Lucerne, Switzerland
e-mail: klaus.mathis@unilu.ch

K. Mathis (ed.), *Efficiency, Sustainability, and Justice to Future Generations*, Law and Philosophy Library 98, DOI 10.1007/978-94-007-1869-2_5,

1.1 The Problem of the Lack of Reciprocity

A contract-theory justification of intergenerational justice is confronted by the following question: why should existing generations take on any obligations towards future generations? What can future generations offer us – other than their undying gratitude, or fond remembrance of us – in return for fulfilling any obligations that we undertake on their account? Echoing Robert L. Heilbroner, we might ask: what has posterity ever done for me?[1]

What is more, future generations unlike people living contemporaneously lack any *potential to pose a threat.* In classic contract theory this is what drives individuals, out of rational self-interest, to adopt an ordered society in place of the unregulated state of nature.[2] Therefore the main problem is rooted in the *non-simultaneity* and the consequent *lack of reciprocity* between present and future generations. Brian Barry draws the succinct conclusion:

> [I]f justice equals mutual advantage then there can be no justice between generations.[3]

This has to be qualified in one respect, however: reciprocity is only missing between remote generations. It certainly exists between directly successive generations, as we know from the figure of the intergenerational contract, described for instance by Otfried Höffe as a phase-shifted exchange that has been fostered from generation to generation for all time: parents care for their children, who later reciprocate by supporting their parents. In a continuous cycle, the process is repeated between the children and the children's children, and so on ad infinitum.[4]

The extrapolation of this exchange model to remote generations can be further underpinned with the following thesis: by ignoring remote generations and making provision solely for the next generation, present generations might leave remote generations even better off than by adhering to intergenerational moral principles.[5] However, this thesis might not hold true for long-term processes of nature degradation, climate change, or hazards that will only become acute in the distant future (known as "sleepers").[6] There are many cases in which acting for the benefit of the next generation or the one after that may have positive or at least neutral impacts, yet be very bad indeed for later generations.[7] Hence the exchange model can only provide very limited justification for obligations to future generations.

In his contract-theory justification of just institutions, John Rawls finds himself similarly confronted by the problem of justice between generations and devotes a

---

[1] Heilbroner, pp. 191 ff.

[2] Cf. Caspar, pp. 92 f.

[3] Barry, *Theories*, p. 180.

[4] Höffe, 'Verteilung', pp. 173 f.; id., *Moral*, pp. 182 f.; id., *Politische Gerechtigkeit,* p. 426.

[5] Cf. Birnbacher, 'Fernstenliebe', p. 30.

[6] Cf. Barry, 'Justice', p. 268.

[7] Barry, *Theories*, p. 193.

chapter of his *Theory of Justice* (§ 44) to the question.[8] Even though this chapter – as we will see – is not entirely persuasive, Rawls' inclusion of future generations is remarkable all the same because it was far from self-evident at the time.

1.2 The "Just Saving" Principle

In his theory, Rawls discusses justice between the generations from the viewpoint of the "just saving" principle. However, this concentration on the question of just saving strongly constricts his perspective and the force of his argument from the outset. Yet for Rawls, saving fulfils a very important function: its purpose is to establish and maintain a *just system* over time. Once the necessary capital base for this has been achieved, the real saving rate can be reduced to zero:

> Real saving is required only for reasons of justice: that is, to make possible the conditions needed to establish and to preserve a just basic structure over time. Once these conditions are reached and just institutions established, net real saving may fall to zero.[9]

By capital, Rawls means more than just mechanical capital – industrial plant and machinery – but knowledge and cultural accomplishments as well:

> It should be kept in mind here that capital is not only factories and machines, and so on, but also the knowledge and culture, as well as the techniques and skills, that make possible just institutions and the fair value of liberty.[10]

In order to recognize the priority of the just saving principle in Rawls' theory, we must briefly remind ourselves of the two principles of justice – the liberty principle and the difference principle – which are the pivotal result of his theory of justice.[11]

The liberty principle:

> Each person is to have an equal right to the most extensive total system of equal basic liberties compatible with a similar system of liberty for all.[12]

The difference principle:

> Social and economic inequalities are to be arranged so that they are both:
>
> (a) to the greatest benefit of the least advantaged, consistent with the just savings principle, and
> (b) attached to offices and positions open to all under conditions of fair equality of opportunity.[13]

[8] Rawls, *TJ*, § 44, pp. 284 ff.

[9] Rawls, *JF*, § 49.2.

[10] Rawls, *TJ*, § 44, p. 288.

[11] For an exposition and critique of the entire theory see e.g. Mathis, 'Theory of Justice', § 7.

[12] Rawls, *TJ*, § 46, p. 302.

[13] Rawls, *TJ*, § 46, p. 302.

As we see, the just saving principle is cited in the difference principle and is a prior constraint upon it. Rawls thereby brings into play not only *intra*generational but also *inter*generational justice:

> The principle of just saving holds between generations, while the difference principle holds within generations.[14]

So Rawls makes a clear distinction between these two questions; moreover, he expressly refuses to apply the difference principle to the question of justice between generations. Therefore the just saving principle cannot be defined so as to give the least favoured the greatest advantage. For the consequence of such an approach would be that either nothing, or not enough, would be saved.[15] Any translation of the difference principle onto the intergenerational relationship would actually prohibit the earlier generation from abstaining from consumption in order to permit all subsequent generations to consume all the more.[16] For Rawls, that outcome would be unacceptable since it would lead to the stagnation rather than the improvement of social conditions.

2 Justice Between Generations

2.1 The Definition of the Original Position

The question of justice towards future generations does not quite seem to fit into Rawls' theory. The problem begins with the fact that the theory is completely atemporal. First of all there is the fictive original position which is static in its conception. It is discontinuously separated from external reality, which can only be called "later" in a metaphorical sense. But even this reality appears to be static in certain points. The whole lifespan that an individual would pass through is reduced to a position in society. This atemporalization is no accident in Rawls' theory: rational choice of distribution principles presupposes that the possible consequences of this choice are in some way universally comparable.[17]

In order to expand his theory to future generations, Rawls reflects on the original position once again from the perspective of multiple generations, and discusses a number of different variations.[18] Firstly he distinguishes between two alternative versions of the original position, called A and B below:[19]

---

[14]Rawls, *JF*, § 49.2.

[15]Hübner, p. 43.

[16]Reuter, p. 186.

[17]Hübner, p. 42.

[18]Rawls, *TJ*, § 24, pp. 136 ff.

[19]Cf. Birnbacher, 'Rawls', pp. 387 f.

A. In the original position, people of all generations assemble. The veil of ignorance ensures that they do not know which generation they belong to. Every person is guided by self-interest.
B. Only people from one generation assemble. They know that they are contemporaries but not which generation they belong to. Again, every person is guided only by self-interest.

Version A would actually stand as the intergenerational analogue to Rawls' intragenerational original position. Instead of different social positions, the parties represent different generations seeking a fair balance between the advantages and disadvantages attaching to possible positions in time, by agreeing on the just saving principle from an apersonal and atemporal viewpoint.[20]

However, Rawls rejects model A as an unreasonable strain on the imagination. It should also be possible to imagine the original position as real at any time, in his view. He opts for a "present time of entry" interpretation of the original position:

> It is not a gathering of all actual or possible persons. [This] is to stretch fantasy too far; the conception would cease to be a natural guide to intuition. In any case, it is important that the original position be interpreted so that one can at any time adopt its perspective.[21]

Having rejected model A in favour of model B, he analyses its impacts on the choice of the saving principle.[22] Now the problem arises that the parties in the original position, all of whom belong to one generation – even if they do not know which one – have no rational reason to agree a saving principle:

> [T]here is no reason for them to agree to undertake any saving whatever. [. . .] Either earlier generations have saved or they have not; there is nothing the parties can do to affect it.[23]

If the generation represented in the original position is selfish, it has no reason to save for future generations because it does not directly benefit from doing so. Nor does it have any power to alter whether earlier generations have saved or not.

2.2 The Supplementary Motivational Assumption

To ensure that the parties in the original position agree a saving principle at all, Rawls therefore has to rethink the premises. In doing so, he modifies the underlying motivation of the parties in the original position by declaring them to be *representatives of ancestral lines* whose immediate progeny are important to them.

> The parties are regarded as representing family lines, say, with ties of sentiment between successive generations.[24]

[20]Birnbacher, 'Rawls', p. 388.
[21]Rawls, *TJ*, § 24, p. 139.
[22]Rawls, *TJ*, § 44, pp. 291 f.
[23]Rawls, *TJ*, § 44, p. 292.
[24]Rawls, *TJ*, § 44, p. 292.

These altruistic feelings are deemed to extend across at least two generations, so that the interests of succeeding generations would overlap.[25] The idea of picking up on the emotional attachment between successive generations is not unlike John Passmore's "chain of love" between the generations, in which the emotional tie between generations perpetuates itself into the distant future.

> Men do not love their grand-children's grand-children. They cannot love what they do not know. But in loving their grand-children – a love which already carries them a not inconsiderable distance into the future – they hope that those grand-children, too, will have grand-children to love. They are *concerned*, to that degree, about their grand-children's grand-children. [...] By this means there is established a chain of love and concern running throughout the remote future.[26]

In making the supplementary motivational assumption, Rawls is therefore opting for a variant C, which represents a modified version of model B:[27]

> C. In model B only people from one generation assemble. They in turn know that they are contemporaries but not which generation they belong to. Unlike the people in models A and B, their interests are not purely selfish; as representatives of families, they are also concerned with the welfare of their next of kin.

This retrospective amendment of the anthropological premises is unconvincing for a number of reasons. For one thing, altruistic sentiments towards immediate descendants contradict the assumption of self-interest in the original position. For another, altruistically motivated intergenerational justice is in conflict with egotistically motivated intragenerational justice.[28] Added to that, this altruistic motivation does not embrace remoter future generations. Therefore it is open to the same criticism as Höffe's exchange model: it largely neglects long-term processes of nature degradation, climate change, or risks that will only become acute in the distant future.

Furthermore, it raises the question as to whether this can really be called justice between generations at all. Since original position C only features contemporaries, whose descendants are no more than their "appendages" and cannot agree to anything themselves, really it can only be called justice which bears future generations in mind.[29] This being the case, there would be no obligations *towards* future generations, only towards members of the present generation *with regard to* future generations.

This construction is strongly reminiscent of Kant's argumentation on the question of whether we have obligations towards animals. This he denies, but nevertheless he does not dismiss such obligations out of hand, interpreting them instead as duties to oneself *with regard to* other beings:[30]

---

[25]Rawls, *TJ*, § 22, p. 128.

[26]Passmore, pp. 88 f.

[27]Birnbacher, 'Rawls', p. 388.

[28]Ackerman, p. 225.

[29]Barry, *Theories*, p. 192.

[30]Birnbacher, 'Natur', pp. 117 f.

> A human being can therefore have no duty to any beings other than human beings; [...] and his supposed duty to other beings is only a duty to himself. He is led to this misunderstanding by mistaking his duty *with regard to* other beings for a duty *to* those beings.[31]

In consequence, the solution with the additional motivational assumption has the drawback that future generations are not fully recognized members of the moral community. Their interests are represented in the original position only by others concerned for their welfare. They, in turn, are not bound to the decision reached in the original position since they were not present when it was agreed.

All this gives rise to yet another problem: even if generation *k* decides on a saving rate, it is far from guaranteed that remote future generations will stand to benefit. That will only happen if subsequent generations fall into line with their chosen savings principle.[32] Even if many generations adhere to the savings principle, it only takes a single *spoilsport generation* to deplete the capital stock and wipe out all the efforts of all previous generations. The most that can be said is that savings built up today put the next generation in a better position, which enables that generation likewise to save for the next generation's benefit.[33] This criticism could be swept aside, however, by contending that each generation can undertake to meet in the original position and decide on the correct savings principle. In that case, it would be bound by the decision reached.

Finally, from the methodological perspective it should be noted that the supplementary motivational assumption contradicts the fundamental tenet of the theory, according to which the principles of justice should be derived from a unified concept.[34] Barry sees Rawls' solution as a blatant trick:

> The only justification offered for the 'motivational assumption' is that it enables Rawls to derive obligations to future generations. But surely this is a little too easy, like a conjurer putting a rabbit in a hat, taking it out again and expecting a round of applause.[35]

The postulate of altruistic interest within the original position therefore compromises the whole systematic derivation from contract theory, according to which any such altruism should result specifically from the combination of the self-interest premise with the structural conditions of the original position.[36] Essentially, the device of the veil of ignorance is intended to help the self-interested person to choose universal principles from an impartial perspective, and consequently to decide upon just institutions. *With the supplementary motivational assumption relating to justice between generations, Rawls therefore undermines his own methodological approach.*

---

[31] Kant, *Methaphysik*, p. 442; English quotation from id., *Metaphysics (trans.)*, p. 192.

[32] Barry, 'Justice', p. 276.

[33] Barry, 'Justice', pp. 276 f.

[34] Höffe, 'Verteilung', p. 179.

[35] Barry, 'Justice', p. 279.

[36] Birnbacher, 'Rawls', p. 393.

2.3 Rawls' Later Proposed Solution

In "The Basic Structure as Subject" (1978) Rawls makes a new suggestion for determining the just saving principle. He gives up his somewhat unconvincing solution relying on the motivational assumption, and now favours model B, in which the parties know that they belong to the same generation, but not which one. Rawls then tries to demonstrate the arguments with which the parties in the original position should decide on the just saving principle.

To determine the just saving rate, he proposes the following line of thought, acknowledged as based on an idea of Thomas Nagel and Derek Parfit and also expounded by Jane English.[37]

> Rather than imagine a (hypothetical and non-historical) direct agreement between all generations, the parties can be required to agree to a savings principle subject to the further condition that they must want all *previous* generations to have followed it. Thus the correct principle is that which the members of any generation (and so all generations) would adopt as the one their generation is to follow and as the principle they would want preceding generations to have followed (and later generations to follow), no matter how far back (or forward) in time.[38]

So the parties in the original position would choose the savings principle that they would wish past generations to have followed and future generations also to follow. In "Political Liberalism" (1993) Rawls repeats this new formulation almost verbatim.[39] Finally, in "Justice as Fairness: A Restatement" (2001) Rawls restates the method for determining the just saving principle:

> They are to ask themselves how much (what fraction of the social product) they are prepared to save at each level of wealth as society advances, should all previous generations have followed the same schedule.[40]

Following this idea, in the original position one can compare various saving rates by comparing the costs and benefits accruing to the generation in the original position as a result of previous generations' adherence to any given saving rate. This rule suggests that a certain amount of saving is beneficial for all generations (except the first).[41] How the just saving principle is determined remains very vague, however. In contrast to the great precision of the difference principle, the reflections on the question of justice between generations arrive at no more than a "seemingly fair outcome", the exact content of which ultimately remains uncertain.[42]

That aside, even the later solution proposed by Rawls fails to resolve the real problem that the generation represented in the original position has no reason to

---

[37] English, p. 98.

[38] Rawls, 'Structure', pp. 58 f.

[39] Rawls, *Liberalism*, p. 274 and p. 20, footnote 22.

[40] Rawls, *JF*, § 49.3.

[41] Mulgan, p. 42.

[42] Hübner, p. 44.

save for future generations. Admittedly the generation in the original position recognizes that this would be the correct rule for determining the just saving principle. Even so, the problem remains that the ones doing the saving will not be the beneficiaries of that saving. In other words, there is an *n-generations prisoners' dilemma.* Generation *k*, which is supposed to decide the just saving principle behind the veil of ignorance, would only be willing to save if previous generations had also saved on a similar scale. But such an agreement *between* different generations is not possible in the model B original position, because only one particular generation *k* is represented there.[43]

2.4 Hume's Conditions of Justice Versus Kant's Universalization

Brian Barry demonstrates, in his article "Circumstances of Justice and Future Generations" (1978), that the question of justice between generations draws attention to a fundamental construction problem in "A Theory of Justice": Rawls' attempt to combine the theories of Hume and Kant leads to certain problems, due to the strongly conflicting nature of these theories. Barry suggests that Rawls adopts Hume's premises but draws Kantian conclusions:

> One is to say that the premises are Hume. This suggestion fits with the fact that the circumstances of justice, which are distinctively Humean, turn up in the premises, while the distinctively Kantian notion that justice has nothing to do with happiness turns up in the conclusions.[44]

According to Hume, justice is an artificial virtue which is dependent on human convention and is only relevant in certain circumstances. As he asserts in "A Treatise of Human Nature" (1739/40):

> [J]ustice takes its rise from human conventions; and [...] these are intended as a remedy to some inconveniences, which proceed from the concurrence of certain *qualities* of the human mind with the *situation* of external objects. The qualities of the mind are *selfishness* and *limited generosity*: And the situation of external objects is their *easy change*, joined to their *scarcity* in comparison of the wants and desires of men.[45]

In "An Enquiry Concerning the Principles of Morals" (1751) Hume takes up this idea once again and points out that, unless very specific conditions prevailed, justice would be virtually useless:

> Produce extreme abundance or extreme necessity: Implant in the human breast perfect moderation and humanity, or perfect rapaciousness and malice: By rendering justice totally *useless*, you thereby totally destroy its essence, and suspend its obligation upon mankind. The common situation of society is a medium amidst all these extremes.[46]

---

[43] Barry, 'Justice', p. 278.

[44] Barry, 'Future Generations', pp. 228 f.

[45] Hume, *Treatise*, Part II, Sect. II.

[46] Hume, *Enquiry*, Part I, Sect. III.

In summary, Hume's conditions of justice can be described as follows: people live together who are approximately similar, do not have all interests in common, and are endowed with limited psychological, emotional and intellectual capacities. On the one hand they exhibit *limited selfishness* and generosity, so that conflicting demands arise; on the other hand, there is *moderate scarcity* of goods. For if goods were available in abundance, it would tend to nullify the obligation to distribute goods with due consideration to others; and, vice versa, if conditions of extreme, life-threatening scarcity prevailed, any obligation to consider others would be unreasonable. Where obligations can no longer be justified, according to Hume, justice becomes irrelevant.[47]

A similar approach to Hume's is that pursued by H. L. A. Hart in the chapter on "The Minimum Content of Natural Law" in his major work "The Concept of Law" (1961).[48] Hart makes the assumption that no moral or legal code can be put in place unless certain physical, mental or economic conditions are satisfied. He talks of "causal connections" between these matters and the emergence of moral and legal systems. Among the prerequisites he mentions are "human vulnerability", "approximate equality", "limited altruism", "limited resources" and "limited understanding and strength of will".[49]

In his theory, Rawls sums up Hume's conditions of justice as moderate scarcity and mutual disinterest.[50] Accordingly, people in the original position take decisions essentially in pursuit of their own interests and in a situation of relative scarcity.[51] In introducing the veil of ignorance, Rawls combines the Humean concept of justice as rational cooperation with the Kantian idea of justice as a universal, hypothetical consensus.

Rawls' resistance to extending the veil of ignorance to knowledge about which generation one belongs to might be rooted in a reluctance to distance himself from the Humean conditions of justice. This would mean departing from the principle that justice rests on mutual advantage.[52] However, the later rule proposed by Rawls is equally problematic in this respect. It applies the categorical imperative to the question of just saving. The outcome could be Kant's, but Rawls starts from Humean premises: *although the parties may recognize the correct moral rule, it is not in their interests to obey it.*

The bridging element between the conflicting theories of Hume and Kant is the veil of ignorance, which ensures that even selfish people adopt an impartial viewpoint and opt for just institutions. *Therefore to establish impartiality from a temporal point of view, there is no alternative but to draw the veil of ignorance over which generation people belong to.* As a consequence, however, the Kantian element is strengthened at the expense of the Humean.

---

[47] Tschentscher, pp. 79 f.

[48] As cited explicitly by Rawls in *TJ*, § 22, p. 126, footnote 3.

[49] Hart, pp. 193 ff.

[50] Rawls, *TJ*, § 22, p. 128.

[51] Kleger, p. 181.

[52] Barry, *Theories*, p. 195.

2.5 Extending the Veil of Ignorance to Membership of a Generation

Rawls himself states that if the veil of ignorance were extended to membership of a particular generation, the motivational assumption would be superfluous.[53] Could model B not be modified so as to retain the idea, favoured by Rawls, that the parties in the original position are contemporaries ("present time of entry interpretation") yet make a decision that is temporally impartial? D. Clayton Hubin has proposed the following solution:[54]

> D. Only people from one generation assemble. But they do not know they are contemporaries, nor which generation they belong to. Every person is guided by self-interest.

This would certainly be a way of obscuring which particular generation people would belong to. All the same, it is a somewhat strange suggestion that parties who meet face to face should be kept in the dark about their status as contemporaries, because people who meet face to face usually *are* contemporaries. This objection appears to be addressed by Otfried Höffe's sophisticated proposal:[55]

> E. The parties in the original position know that they are only contemporaries in an ideal sense – as an operative means of representing the viewpoint on justice – but in actual fact they belong to different generations. Every person takes self-interest as their guide.

Höffe's suggestion would certainly work, too, but is even more artificial than Hubin's. Therefore the question is whether it is not better to resort to the simpler model A, after all.

Rawls opposed model A, firstly because he believed that it taxed the imagination unduly and secondly because he wanted it to be possible to imagine the original position in reality at any time. But in the first place, abstraction from the time in which one will live ought to be no harder than abstraction from the abilities and position in society that one will have.[56] And as to the second point, Rawls has obviously confused two different things: the idea that one can imagine the original position as real at any time, and the idea that the people behind the veil of ignorance must be contemporaries.[57]

The gathering in the original position can be imagined as real at any time, even if the veil of ignorance conceals which particular generation one belongs to. In fact, as a thought experiment, the discussion in the original position can even be conducted by one person alone: since the parties base their decisions on the same considerations and take decisions unanimously, the agreement reached in the original position can equally well be seen as that of one randomly chosen participant reaching the

---

[53]Rawls, *TJ*, § 44, pp. 291 f.

[54]Hubin, p. 75.

[55]Höffe, 'Verteilung', p. 179 (trans. here from the German for convenience).

[56]Visser 't Hooft, p. 74.

[57]Cf. Ott, p. 97.

same agreement alone. The advantage of B over A lies only in greater ease of intelligibility, and not any basis in substance.[58] Thus, there is nothing whatsoever to gainsay the choice of model A, in which the original position is imagined as a gathering of people from *all generations*, albeit that they do not know which generation they belong to.

As a counterbalance, however, the parties in the original position should be privy to *general ecological understanding* about the relationship between man and nature so that they can make rational decisions about issues of justice between the generations.[59] This knowledge is necessary most of all because the question of what should actually be "saved" is much more complex than in Rawls' presentation. The level of the saving rate or the capital stock are far from the only definitive factors. The crux of the matter is what types of capital should be saved, or should not be destroyed. As we will see, what is known as "natural capital" is of central importance.

3 What Should Actually Be "Saved"?

The problem of what capital stock should be saved, and what this means in concrete terms, is a major bone of contention in the current sustainability debate. The critical question is the "sustainability of what?", following on from the famous "equality of what?" debate prompted by Amartya Sen.[60]

In the literature, a fundamental distinction is made between *weak sustainability* and *strong sustainability*. The weak sustainability concept has its origins in traditional, neoclassical "Environmental and Resource Economics". Strong sustainability, on the other hand, is an idea formulated by the proponents of a newer, interdisciplinary strand of research known as "Ecological Economics".

3.1 Weak Sustainability

The concept of weak sustainability demands the *preservation of the entire capital stock*. Put simply, living off capital assets is not allowed. The capital stock K can be subdivided into natural capital K_N, mechanical capital K_M and human capital K_H. Therefore: $K = K_N + K_M + K_H = const.$ Natural capital includes raw materials, energy, nature reserves and landscapes, and ecosystems generally. Mechanical capital consists of capital in the narrow sense, e.g. tools, machinery, buildings or transport infrastructure. Human capital refers to human knowledge, skills and abilities.[61]

[58]Birnbacher, 'Rawls', p. 388.

[59]Cf. Ott, pp. 96 f.

[60]Sen.

[61]Weikard, pp. 54 f.

The different forms of capital *can, in principle, be freely substituted* (empirical interpretation) – and *may indeed be substituted* (normative interpretation). It follows that natural capital may be consumed if mechanical and/or human capital are commensurately increased. The rents from the consumption of natural resources must therefore be reinvested (known as the Hartwick rule).[62] The concept of weak sustainability tends to be combined with optimism about technology: an assumption is made that *technological progress* will overcome resource constraints.[63] This viewpoint is predominant among traditional economists.

The proponents of *ecological economics* are less optimistic in this respect. They doubt the ability of the market to bring about adequate substitution and innovation effects by means of price adjustment mechanisms.[64] In any case, they consider natural capital to be fundamentally unsubstitutable: they argue that ecological systems are not simply production factors that can be replaced with others; they are the very foundation of all life and economic endeavour. On this rationale, a concept of strong sustainability is required.[65]

3.2 Strong Sustainability

According to the concept of strong sustainability, therefore, what should be preserved is not just the entire capital stock K, but specifically also the *stock of natural capital*: $K_N = const.$[66] A loss of natural capital cannot be balanced out by a greater stock of mechanical and/or human capital. Now this requirement to preserve natural capital can be interpreted in a variety of ways:

1. Value-based preservation of natural capital;
2. Physical preservation of natural capital;
3. Physical preservation of critical natural capital.

Under *value-based preservation of natural capital*, within the category of natural capital all elements are freely substitutable. Non-renewable resources such as coal may be consumed as long as rents are reinvested, e.g. in renewable energies. So the possibility of substitution weakens the strong concept of sustainability. It can also produce questionable results: it is hardly plausible to claim that whereas additional mechanical capital can never make up for enlargement of the hole in the ozone layer, increasing the whale stocks in the world's oceans somehow can.[67]

---

[62]Weikard, pp. 55 f.

[63]Neumayer, p. 23.

[64]Nutzinger, p. 71.

[65]Costanza and Daly, pp. 2 f.

[66]Weikard, p. 55.

[67]Neumayer, p. 25.

The above reflections show that substitution even within natural capital cannot necessarily be reconciled with a concept of strong sustainability. If we want to rule out substitution, we have to argue for physical preservation of natural capital. An extreme form of sustainability would insist on physical preservation of the *entire stock of natural capital* in situ. But this would forbid absolutely any further changes to nature, implying an ultra-conservative and more or less life-inhibiting attitude.

A more sophisticated view of strong sustainability is that only certain parts of the natural capital must be physically preserved, namely those which are effectively unsubstitutable. This is known as *critical natural capital.* Use of non-renewable resources would be permissible if part of the rent is invested in renewable energies. Destruction of the Amazon rainforest, on the other hand, would be unacceptable and the diversity of animal and plant species would need to be conserved. The general principle is to ensure that the *functions of nature* remain intact.[68]

A concept of strong sustainability ought to satisfy the following three conditions *cumulatively:* value-based preservation of the entire capital stock (minimum requirement for weak sustainability); value-based preservation of natural capital (supplementary requirement for strong sustainability); and, since even elements of natural capital are not fully substitutable, *physical preservation of critical natural capital* is also a necessity. This physical preservation of critical natural capital forms the *pièce de résistance* of strong sustainability, because it is the only way to guarantee the vital life-supporting functions of nature now and for future generations.[69]

Finally, it is worth mentioning that what capital stock to preserve is not the only significant issue: there is also the question of what *hazards* we leave behind for future generations. Particular thought might be given to nuclear risks, or to the problem of overpopulation.[70]

4 Implications

Rawls' treatment of the question of intergenerational justice is confined to the aspect of "just saving". He discusses saving in relation to the entire capital stock of a society; natural capital is not treated as a separate category. Here Rawls makes a tacit assumption that the various types of capital are substitutable. Thus his arguments are founded on the *paradigm of weak sustainability* which does not adequately account for ecological aspects.

However, there is no compelling need to focus on a concept of weak sustainability based on the structural conditions of the original position. This narrowing of perspective or, perhaps, unfounded preconception has to be ascribed to Rawls

[68]Neumayer, p. 25.

[69]The reference to physical dimensions has the further advantage of obviating the valuation and aggregation problem that rears its head in relation to monetary dimensions.

[70]Cf. Bickham, p. 173.

himself. Another important aspect of sustainability, namely the question of what hazards we might acceptably leave behind for later generations, is also missing from Rawls' discussion. Nevertheless it would be very easy to expand his theory, by incorporating these aspects, in the direction of *strong sustainability.*

Nor, as we have seen, is there any objection to *extending the veil of ignorance to one's membership of a particular generation*. The agreement in the original position can be imagined as a thought experiment at any time, even if the veil of ignorance obscures which generation one belongs to. If individuals were not biased towards any particular era, what issues might they consider in the original position? Doubtless they would ask themselves what essential life-sustaining resources are necessary to safeguard the basic prerequisites for a just society for all time.

Bibliography

Ackerman, Bruce A., *Social Justice in the Liberal State* (New Haven and London, 1980)

Barry, Brian, 'Circumstances of Justice and Future Generations', in Richard I. Sikora and Brian Barry (eds.), *Obligations to Future Generations* (Philadelphia, 1978), pp. 204 ff. (cited as: 'Future Generations')

Barry, Brian, 'Justice Between Generations', in Peter M. S. Hacker and Joseph Raz (eds.), *Law, Morality, and Society* (Oxford, 1977), pp. 268 ff. (cited as: 'Justice')

Barry, Brian, *Theories of Justice* (London et al., 1989; cited as: *Theories*)

Bickham, Stephen, 'Future Generations and Contemporary Ethical Theory', in *Journal of Value Inquiry*, Vol. 15 (1981), pp. 169 ff.

Birnbacher, Dieter, '"Fernstenliebe" oder Was motiviert uns, für die Zukunft Vorsorge zu treffen?', in Ralf Döring and Michael Rühs (eds.), *Ökonomische Rationalität und praktische Vernunft. Gerechtigkeit, Ökologische Ökonomie und Naturschutz* (Würzburg, 2004), pp. 21 ff. (cited as: 'Fernstenliebe')

Birnbacher, Dieter, 'Rawls' "Theorie der Gerechtigkeit" und das Problem der Gerechtigkeit zwischen den Generationen', in *Zeitschrift für philosophische Forschung*, Vol. 31 (1977), pp. 385 ff. (cited as: 'Rawls')

Birnbacher, Dieter, 'Sind wir für die Natur verantwortlich?', in id. (ed.), *Ökologie und Ethik* (Stuttgart, 1980), pp. 103 ff. (cited as: 'Natur')

Caspar, Johannes, 'Generationen-Gerechtigkeit und moderner Rechtsstaat. Eine Analyse rechtlicher Beziehungen innerhalb der Zeit', in Dieter Birnbacher and Gerd Brudermüller (eds.), *Zukunftsverantwortung und Generationensolidarität* (Würzburg, 2001), pp. 92 f.

Costanza, Robert and Daly, Hermann, 'Towards an Ecological Economics', in *Ecological Modelling*, Vol. 38 (1987), pp. 1 ff.

English, Jane, 'Justice Between Generations', in *Philosophical Studies*, Vol. 31 (1977), pp. 91 ff.

Hart, Herbert Lionel Adolphus, *The Concept of Law* (2nd edn., Oxford, 1994)

Heilbroner, Robert L., 'What Has Posterity Ever Done for Me?', in Ernest Partridge (ed.), *Responsibilities to Future Generations. Environmental Ethics* (Buffalo, 1981), pp. 191 ff.

Höffe, Otfried, *Moral als Preis der Moderne. Ein Versuch über Wissenschaft, Technik und Umwelt* (Frankfurt a.M., 1993; cited as: *Moral*)

Höffe, Otfried, *Politische Gerechtigkeit. Grundlegung einer kritischen Philosophie von Recht und Staat* (3rd edn., Frankfurt a.M., 2002; cited as: *Politische Gerechtigkeit*)

Höffe, Otfried, 'Zur Gerechtigkeit der Verteilung', in id. (ed.), *John Rawls – eine Theorie der Gerechtigkeit* (Berlin, 1998), pp. 169 ff. (cited as: 'Verteilung')

Hubin, D. Clayton, 'Justice and Future Generations', in *Philosophy and Public Affairs*, Vol. 6 (1976/77), pp. 70 ff.

Hübner, Dietmar, 'Justice over Time. Zum Problem der Gerechtigkeit zwischen den Generationen', in *Jahrbuch für Wissenschaft und Ethik*, Vol. 6 (2001), pp. 39 ff.

Hume, David, *A Treatise of Human Nature,* Book III (London, 1739/40; cited as: *Treatise*)

Hume, David, *An Enquiry Concerning the Principles of Morals* (London, 1751; cited as: *Enquiry*)

Kant, Immanuel, *Metaphysik der Sitten, Akademieausgabe*, Vol. VI (Berlin, 1968; cited as: *Metaphysik*)

Kant, Immanuel, *The Metaphysics of Morals*, trans. Mary Gregory (Cambridge, 1996; cited as: *Metaphysics (trans.)*)

Kleger, Heinz, 'Gerechtigkeit zwischen Generationen', in Peter Paul Müller-Schmid (ed.), *Archiv für Rechts- und Sozialphilosophie (ARSP), Beiheft 26 "Begründung der Menschenrechte"* (Stuttgart, 1986), pp. 147 ff.

Mathis, Klaus, 'John Rawls' Theory of Justice', in id. (ed.), *Efficiency Instead of Justice? Searching for the Philosophical Foundations of the Economic Analysis of Law* (Dordrecht, 2009), § 7 (= 'Die Theorie der Gerechtigkeit von John Rawls', in *Effizienz statt Gerechtigkeit? Auf der Suche nach den philosophischen Grundlagen der Ökonomischen Analyse des Rechts* (3rd edn., Berlin, 2009, § 7); cited as: 'Theory of Justice')

Mulgan, Tim, *Future People. A Moderate Consequentialist Account of our Obligations to Future Generations* (Oxford and New York, 2006)

Neumayer, Eric, *Weak Versus Strong Sustainability. Exploring the Limits of Two Opposing Paradigms* (2nd edn., Cheltenham and Northampton, 2003)

Nutzinger, Hans G., 'Langzeitverantwortung im Umweltstaat aus ökonomischer Sicht', in Carl Friedrich Gethmann et al. (eds.), *Langzeitverantwortung im Umweltstaat* (Bonn, 1993), pp. 42 ff.

Ott, Konrad, 'Essential Components of Future Ethics', in Ralf Döring and Michael Rühs (eds.), *Ökonomische Rationalität und praktische Vernunft. Gerechtigkeit, Ökologische Ökonomie und Naturschutz* (Würzburg, 2004), pp. 83 ff.

Passmore, John, *Man's Responsibility for Nature. Ecological Problems and Western Traditions* (London, 1974)

Rawls, John, *A Theory of Justice* (Cambridge, 1971; cited as: *TJ)*

Rawls, John, *Justice as Fairness. A Restatement*, ed. Erin Kelly (Cambridge and London, 2001; cited as *JF*)

Rawls, John, *Political Liberalism* (New York, 1993; cited as: *Liberalism*)

Rawls, John, 'The Basic Structure as Subject', in Alvin I. Goldman and Jaegwon Kim (eds.), *Values and Morals. Essays in Honor of William Frankena, Charles Stevenson, and Richard Brandt* (Dordrecht, 1978), pp. 47 ff. (cited as: 'Structure')

Reuter, Hans-Richard, 'Der "Generationenvertrag". Zur ethischen Problematik einer sozialpolitischen Leitvorstellung', in Karl Gabriel and Hermann-Josef Grosse Kracht (eds.), *Brauchen wir einen neuen Gesellschaftsvertrag?* (Wiesbaden, 2005), pp. 171 ff.

Sen, Amartya, *Equality of What? The Tanner Lecture on Human Values* (Cambridge, 1980)

Tschentscher, Axel, *Prozedurale Theorien der Gerechtigkeit. Rationales Entscheiden, Diskursethik, Recht* (Baden-Baden, 2000)

Visser 't Hooft, Hendrik Ph., *Justice to Future Generations and the Environment* (Dordrecht, 1999)

Weikard, Hans-Peter, *Wahlfreiheit für zukünftige Generationen. Neue Grundlagen für eine Ressourcenökonomik* (Marburg, 1999)

What Is It Like to Be Unborn?

Our Common Fate with Future Generations

Malte-Christian Gruber

> *Justice remains, is yet, to come, à* venir, *it has an, it is à*-venir, *the very dimension of events irreducibly to come. It will always have it, this à*-venir, *and always has. Perhaps it is for this reason that justice, insofar as it is not only a juridical or political concept, opens up for* l'avenir *the transformation, the recasting or refounding of law and politics.*[1]

The International Year of Biodiversity presents an opportunity to revisit the underlying causes of the relentless loss of biological as well as cultural diversity and to reconsider the anthropocentric framing of environmental conservation. As long as economic and juridical perspectives do not question the stance that culture and nature – and especially biodiversity – are to be safeguarded for the sake of present generations only, they will miss the aim of sustainability. To achieve sustainability, jurisprudence must look to the future as well. By embracing the fiction of future generations as communities of fate, the law will be able to reach beyond the borders of its own limited system of concepts and realize that biodiversity does not serve solely economic utilization and social communication, but in addition carries a tangible (ecocentric) weight for physical survival in the future.

1 Our Common Future: Biodiversity and Biotechnology

The General Assembly of the United Nations declared the year 2010 to be the International Year of Biodiversity, calling for the global conservation of vital ecosystems and the protection of endangered flora and fauna. This was borne out of the realization that biodiversity is an existential prerequisite for human life and

[1] Derrida, pp. 969 ff.

M.-C. Gruber (✉)
Institute of Economic Law, Goethe University Frankfurt, Grüneburgplatz 1, D-60629 Frankfurt am Main, Germany
e-mail: gruber@jur.uni-frankfurt.de

K. Mathis (ed.), *Efficiency, Sustainability, and Justice to Future Generations*, Law and Philosophy Library 98, DOI 10.1007/978-94-007-1869-2_6,

wellbeing and this applies to the material, social and cultural aspects.[2] Through the anthropocentric framing of biodiversity conservation, which is clearly geared to human benefit, our awareness is drawn to our dependence on nature which cannot be overcome at all, not even through technological means. At the same time one must not forget that nature's genetic diversity is indispensable for the technical progress of mankind. This allows us to assign the economic value to the genetic resources of fauna and flora, and also to humankind.

This special appreciation of biodiversity is the source of a complex and global conflict over economic growth and sustainable development, biodiversity conservation and use and also over intellectual property rights to traditional knowledge. The stakeholders and their interests are numerous and varied. This conflict can by no means be reduced to a simplified two-party conflict with contending claims. It appears more appropriate to reconstruct the conflict as a "trialogue" between the realms of biodiversity, biotechnology and traditional knowledge.[3] In other words, it is a question of taking an all-encompassing view of the difficult relationship between economic development, technological progress and biological as well as cultural diversity.

1.1 Interdependencies

One could be tempted to describe the latter problem as a "local" interest which must try to resist the "global" superiority of economy and biotechnology. Yet the matter is not that simple. Just as "global" issues increasingly impact on local conditions, global problems can only be managed at local level. The fact that the "global" is increasingly becoming "local" and the "local" "global" is verbalized as follows by Bruno Latour:

> No place dominates enough to be global and no place is self-contained enough to be local.[4]

On the one hand, it is this technological and economic dependence on genetic resources that gives rise to the worldwide interest in the conservation of biodiversity. While on the other hand, it is precisely these biotechnologies which provide new technological methods for the conservation of living diversity in the various regions of the world.

For example, in 1970 Costa Rica began to gradually convert its farming to more ecologically sustainable methods for the use of its still ample natural resources.[5] However, today it is no longer just a matter of moving away from the monoculture of banana and coffee plantations, to reforestation and converting the (national) economy to low-impact eco-tourism, but also to conduct research into technological solutions to "save, know and use" sustainably the country's

---

[2]See UNESCO, p. 2.

[3]King, pp. 428 ff.

[4]Latour, *Reassembling the Social*, p. 204.

[5]Gámez, pp. 78 f.

biodiversity.[6] The Instituto Nacional de Biodiversidad (INBio), a research institution founded in 1989 for the purpose of scientifically researching and evaluating biodiversity, is now also responsible for its sustainable use. To achieve this, this non-governmental organization is working in cooperation with both public institutions and the private sector. The institute has and continues to form, among other things, strategic alliances with multinational commercial enterprises from the biotechnology and pharmaceutical sector. Using individual research collaborative agreements (RCAs) as a base, the cooperating partners make concerted efforts to gather knowledge regarding sustainable utilization as well as the possible commercial use of these biological resources for profit.[7]

This search for innovative, valuable products, termed "bioprospecting", results primarily in an *economic* valuation of biodiversity. This is indeed important in order to apply an – economically relevant – "countervalue" to biological resources compared to other goods in the global economy. However, such an economic valuation can carry the danger of an exclusive commercialization of nature. To avoid this danger, contract parties are seeking to shape the RCAs in line with the principles of access and benefit sharing (ABS) so that access to natural resources remains limited in space and time. Furthermore, a share of the resulting technical developments and profits, will be assigned to benefit local communities and support the on-site conservation.[8]

However, the harmonious wording of these agreements cannot conceal that, at least from the perspective of the financing enterprises, as per usual only those natural resources are of interest that can be utilized, in the foreseeable future, as tradable products such as: pharmaceuticals, agricultural pest control agents, crop or ornamental plant varieties. The basic economic orientation to tangible usefulness[9] can therefore only partly be aligned with the conservation of biodiversity, despite many statements of intent to the contrary. Ultimately, this level of conservation is granted only to that natural diversity that can be utilized commercially – and in the not too distant future. Doubtlessly the gain generated from the natural, raw material must reach a profitable level in relation to the capital invested. In view of the wide range of still undiscovered reserves of "Green Gold"[10] this may be possible without difficulty, but a good price cannot be achieved for every natural resource on the market of patentable natural assets.

1.2 Conflicts

A closer look reveals that the protective measures for biological resources are applied very selectively. In bioprospecting, biological materials which have already

[6]Gámez, p. 79.
[7]Gámez, pp. 81 ff.; Medaglia, pp. 243 ff.
[8]Quezada, pp. 37 ff.; see also Stoll, pp. 3 ff.
[9]See Mathis, *Efficiency*, pp. 7 ff.
[10]See e.g. Wullweber, pp. 3 ff.

proven to be useful and therefore appear profitable are predominantly collected and explored with regard to their genetic information. To minimize the research and development costs for pharmaceutical and agricultural products it is particularly worthwhile to have access to the already existing, traditional knowledge of the possible uses for these natural resources. Therefore, research activities concentrate particularly on those natural substances that have already been utilized for a long time by the local people, for instance for medicinal purposes or for nourishment.

The situation appears paradoxical: the international research enterprises are drawing on the traditional knowledge of small parts of the population, decoupled from the world market, in order to identify economically profitable substances. Their innovative capacity of these substances rests on this traditional knowledge which itself was only able to develop and be passed down through the generations because it has been isolated from the world market and modern "world knowledge" till now.[11]

This decoupling from the worldwide modernization developments has not only resulted in the existence of such cultural diversity of knowledge but also in biodiversity itself[12] – this is highlighted by those who consider bioprospecting to be nothing but a special form of biopiracy, namely the exploitation of natural resources using the tools provided by the modern law of "intellectual property":

> Biodiversity has been protected through the flourishing of cultural diversity. Utilizing indigenous knowledge systems, cultures have built decentralized economies and production systems that use and reproduce biodiversity. Monocultures, by contrast, which are produced and reproduced through centralized control, consume biodiversity.[13]

Thus, if the conservation of biodiversity is reliant on the variety of traditional knowledge communities, then the inverse argument allows us to conclude that it cannot be saved with the means of an economy geared to global commercial utilization alone.[14] A sustainable use of natural resources is certainly not achievable by harnessing them for the global market by means of transforming their genetic information and the associated traditional knowledge into "intellectual property". This applies even if there are good intentions to act in accordance with ABS standards. Thus, even if one intends to share the economic benefits arising from the relevant patent and plant variety rights with the original knowledge carriers or channel these benefits to nature conservation, one can only in part evade the accusation of biopiracy. While it may be possible to provide monetary compensation for people and nature, no one can render "restitution in kind" (i.e. restoration of the original situation) in the full sense of the word.

There is no compensation for the resulting removal of natural and cultural resources through scientific exploration from their original evolutionary context and transforming them into information utilizable for modern economies. They are then

[11] Brand, pp. 129 f.; Brand and Görg, pp. 31 f.

[12] Brand, p. 133.

[13] Shiva, p. 72.

[14] Shiva, pp. 72 ff.

in danger of losing their significance for those life-worlds to which they owe their existence. What is more: if rules, based on culture and/or religion, governing access to traditional knowledge are displaced by the modern exploitation logic, then this will damage the entire cultural development of traditional knowledge communities.[15] Once their sources are exploited, it must be feared that no "new" traditional knowledge can emerge.

Similar problems are faced by the conservation of nature itself: one can collect, secure and continue to use in laboratories ("ex situ") a part of its diversity, yet this does not improve the situation of living organisms in the wild ("in situ").[16] Nor can monetary payments really compensate environmental damage which sometimes is caused by the research activities themselves. Apart from the fact that such monetary compensation is usually employed elsewhere, there are no universal rules providing guidance as to how and which protective measures could completely compensate an environmental damage that has already occurred.

1.3 Valuing Biodiversity: A Matter of Justice

For this reason, strenuous efforts are under way to determine the economic value of natural resources more accurately and specifically.[17] But these attempts remain limited by their own premise that nature should benefit humans – and only humans (above all the presently living generation). Since the earliest phase of designing an independent body of environmental law it is well understood that this anthropocentric perspective remains restricted in the end to the economically calculable values assumed by it:

> [T]he only values consistently served are those strongly felt by persons motivated and able to seek a policy analyst's aid – a circumstance likely to exclude values too widely diffused over space, or too incrementally affected over time, to be strongly championed by any single client of a policy analyst; values associated primarily with persons not yet in being (future generations); and values not associated with persons at all (for example, the "rights" of plants or animals).[18]

The quotation above is from an essay published in 1974 by Laurence Tribe, a US American lawyer. Its striking title reads: "Ways Not To Think About Plastic Trees". Taking the example of "plastic trees" which had been used as cost-saving replacement for natural tree-planting along a main road in Los Angeles, he levelled fundamental criticism at the anthropocentric "want-oriented" perspective.[19] He saw the problem in a deeply anchored "ideological bias of the system",[20] the value

[15] Shiva, p. 74; Teubner and Fischer-Lescano, pp. 17 ff.

[16] Chen, p. 52.

[17] See e.g. Richardson and Loomis, pp. 25 ff.

[18] Tribe, p. 1319.

[19] Tribe, pp. 1315 ff., 1329 ff.; see also Gruber, 'Rechte des Lebendigen', pp. 1546 ff.

[20] Tribe, p. 1332.

definition of which is oriented exclusively to individual human interests which, moreover, are treated as interchangeable.[21] In such a reductionist perspective, it becomes unobjectionable to replace the natural vegetation with plastic trees, indeed even to synthesize the whole of nature provided this is of benefit for humans.

> By treating individual human need and desire as the ultimate frame of reference, and by assuming that human goals and ends must be taken as externally "given" (whether physiologically or culturally or both) rather than generated by reason, environmental policy makes a value judgment of enormous significance. And, once that judgment has been made, any claim for the continued existence of threatened wilderness areas or endangered species must rest on the identification of human wants and needs which would be jeopardized by a disputed development. As our capacity increases to satisfy those needs and wants artificially, the claim becomes tenuous indeed.[22]

Today's efforts to conserve biodiversity in the interest of mankind rest on the same premise. Environmental policy considers nature, life and traditional knowledge in an instrumental manner as natural and cultural resources and converts them into interchangeable values. There are certainly good reasons for this: after all, assigning a value to natural resources requires a set of reference values. And the common measure of human benefit – above all the appeal to the own vital interests – can motivate people to advocate ecological concerns more intensively.

To reconstruct nature and culture as economic resources can serve the protection of biological and cultural diversity, yet one must keep the pitfalls of this in sight. The danger lies in setting economic values, utilitarian considerations and anthropocentric fixations upon human self-interest as absolute, and in this way failing to recognize the intrinsic value of nature and culture beyond a purely instrumental rationale. This is the temptation of economistic reductionism: to give present, frequently short-term personal interests priority over sustainable development, to substitute the factual influence of current players for legitimate justification, to measure values exclusively according to quantity instead of quality – in short: to view only the present instead of future, facticity instead of validity and "efficiency instead of justice".[23]

Based on such a standpoint it becomes difficult to align decisions to long-term perspectives, the consequences of which are scarcely calculable today. Impact assessments are doomed to failure because the "present future" is diverging ever more widely from the "future presents".[24] Decisions therefore appear increasingly risky and no longer concern only those who make them. The resulting dangers cannot be fully anticipated, neither with regards to the circle of affected people nor the true extent of it.[25]

---

[21] Tribe, pp. 1330 ff.

[22] Tribe, p. 1326.

[23] Mathis, *Efficiency*, pp. 185 ff.

[24] Luhmann, 'Risiko und Gefahr', pp. 158 ff.; see also id., *Risk*, pp. 1 ff. and pp. 33 ff.

[25] Luhmann, 'Verständigung', pp. 352 f.

This rather narrow point of view cannot simply be put down to an "economization" of nature and culture. Economic valuation as such has its justification where it identifies the economic equivalents of natural and cultural resources and transforms them into cost-benefit analyses. Economic rationality therefore does not imperatively contradict sustainable environmental conservation. Problems only arise because of its mostly unquestioned nexus with the anthropocentric assumption that solely the interests of the current generation must be considered. This premise leads to the further misplaced assumption that the interests of current people can be depicted entirely in the economic medium, i.e. in "monetary" value. The underlying cause of the short-sightedness of some economic perspectives can be found here: in their inadequate reflection regarding the character of the various interests and the circle of the possible carriers of these interests – and not in economic thinking as such, which fulfils the important function as a partial rationality of a societal sub-system.[26]

As long as economic perspectives assume unquestioning that culture and nature and especially biodiversity are to be safeguarded for the sake of mankind alone, they will not be able to overcome the short-term perspective – not even if they apply the interest of these traditional knowledge communities who are hard pressed by the economic powers. It is only a short-term benefit for the indigenous population and not a long-lasting one if they are included legally and financially in patents that are developed on the basis of their knowledge pool. Participation in a modern market development process may be conducive to the wellbeing of present indigenous groups and may help to secure their existing bodies of knowledge. However, at the same time the processes of generating new traditional knowledge and thus the future basis of existence of indigenous cultures are thereby jeopardized.[27]

1.4 Future Justice (1): The Intrinsic Value of Natural and Cultural Resources

In view of these deficiencies, it is important to recognize an intrinsic value of traditional knowledge – not as a simple knowledge inventory, but as a productive knowledge context – which reaches beyond the merely instrumental dimension of knowledge utilization. Sustainability means, from this point of view, to ask anew about the stakeholders involved and their possible rights of dominion: to whom does nature and the understanding of it belong?

Normally, this question is cast merely in terms of the alternative whether natural and immaterial (intellectual) assets are the property of individuals or are in the public domain as common good. Framing the question in that way, however, already presupposes that nature and traditional knowledge are in principle available, quantifiable, exchangeable, purchasable, in short: commercializable. But precisely this

[26]Luhmann, *Wirtschaft*, pp. 43 ff.

[27]Teubner and Fischer-Lescano, p. 21.

is an epistemic trap of property-oriented, especially patent-legal thinking:[28] a priori the further possibility is overlooked that, due to their intrinsic value, natural and cultural assets may not be exploited at all – in any case not exhaustively – no matter whether through commercialization or through "free" public dissemination. Therefore the question should be more accurately: to whom *may* nature and the understanding of it belong and to *what extent*? This of course makes the issue more complicated but also makes it clear that it is not only a matter of who shall have property rights under the aspect of "optimum" (this almost always means "efficient") resource allocation. The core problem lies in the determination of which actors are eligible as possible carriers of rights and to what extent natural and cultural assets must be considered to be non-available.

The efforts of bioprospecting, to explore biological diversity, should therefore not merely be directed at tapping new exploitable resources but in the first instance at identifying the actors and their interests that have to be taken into account. The question of possible further stakeholders, in view of the intrinsic value and non-availability of nature and knowledge can be construed as part of an overarching problem which has become omnipresent in the age of biotechnology: the increasingly difficult delimitation as to whether living beings – as "body", "parts of a body", "body functions" or "body information" – "appertain" to somebody (as property) or already "are" somebody (as part of a person).

2 Custodians of Biological and Cultural Diversity

In Costa Rica it may appear less alarming, if natural resources that were unknown as yet, are explored and tapped for the first time. Because of the small proportion of indigenous people in the population, there is scarcely the suspicion that natural assets encountered there already belong to someone, be it a substance or some traditional knowledge about it.[29] However, when extending the circle of affected local groups, the fact must be realized that at least the rural population must have knowledge regarding traditional forms of utilization of now explored natural resources and would have good reason to insist on its rights "from tradition".[30]

Overall, most of the difficult cases are those in which the local population is affected. The circumstance that indigenous groups most commonly do not know of the rights to natural and cultural resources comparable to modern "ownership" aggravates the problem. They must then be enabled to preserve their common natural and cultural assets without declaring them the property of any individual.[31] It is essential to empower them, both legally and technically, as independent custodians

---

[28] Teubner and Fischer-Lescano, pp. 34 f.

[29] Gámez, p. 83.

[30] Miller, pp. 359 ff.; Kloppenburg and Rodríguez Cervantes, pp. 12 ff.; Rodríguez Cervantes, pp. 140 ff.

[31] Teubner and Fischer-Lescano, pp. 31 ff.

capable of guarding the production conditions of their traditional knowledge from private and also public disposal.

2.1 Rights of Native People and Farmers' Rights

An important step in this direction is to equip them with the necessary technical means and to assist them in independently mapping and designating their communal lands and resources. In Venezuela, for instance, such initiatives are supported by constitutional "Rights of Native People".[32, 33]

There are further activities to provide technical support and cooperation in other Latin American[34] countries as well as in Asia[35] and Africa,[36] where high hopes are pinned on agricultural biotechnology within the framework of public-private partnerships. Here, too, technology is tailored to the specific requirements of the domestic population. At the same time the scientific and economic development of the biotechnology sector is fostered while preserving biological diversity as far as possible.

It is not clear under which conditions diversity is to be preserved and in which circumstances biotechnology helps to attain this goal. Hence there is no lack of criticism levelled against new, partly irreversible dependencies on technology transfer.[37]

Similar to the process by which the indigenous people as traditional knowledge communities become trustees of their common cultural heritage through the influence of biotechnology, the biotechnological transformation in agriculture gives rise to new interest communities: in particular the great number of rural communities for whom the traditional agrarian culture are indispensable for independent food production in these developing countries. They are deemed guarantors of agricultural biodiversity.[38] *Farmers' rights*[39] are to assure, as collective rights, that they remain independent of industrial farming and its new technologies. The latter include by no means only monocultures facilitated by "external" treatment with plant protection products, but also those treated "internally", i.e. genetically modified crops.[40]

---

[32]See Art. 119 ff. of the Constitución de la República Bolivariana de Venezuela (1999), available at <http://pdba.georgetown.edu/Constitutions/Venezuela/ven1999.html> (visited 2 September 2010).

[33]Zent and Zent, pp. 103 ff.

[34]See e.g. Rosenthal, pp. 373 ff.

[35]Glenn D. Stone, pp. 207 ff.

[36]Wambugu, pp. 174 ff.; Khush, pp. 179 ff.

[37]Busch, pp. 202 ff.; Egziabher, pp. 158 ff.

[38]See Decision V/5 of the Conference of the Parties to the Convention on Biological Diversity (CBD), available at: <http://www.cbd.int/decision/cop/?id=7147> (visited 2 September 2010).

[39]See Art. 9 of the International Treaty on Plant Genetic Resources for Food and Agriculture (FAO International Treaty), available at: <http://www.fao.org/legal/treaties/033t-e.htm> (visited September 2, 2010).

[40]Egziabher, pp. 170 ff.

Once genetically modified seed is sown, it cannot be monitored fully how transgenic plants propagate and mix with other crop and wild plants. The impacts of "Green Genetic Engineering" are drastic compared to natural evolution, and just as irreversible.

2.2 *Future Justice (2): Rights of Biosocial Communities of Fate*

The use of biotechnologies can albeit be useful and sometimes even indispensable in the preservation of biological and cultural diversity for the future. But there are pitfalls too: irreversible and disadvantageous changes to the future. It is therefore all the more important to involve all stakeholders, and particularly the above-mentioned local groups, in decisions that determine the future.

These people are even more affected by biotechnological interventions which do not remain restricted to natural and cultural resources but focus directly on their own genetic constitution. There are in fact already extensive attempts – in Costa Rica amongst others[41] – to gather the genetic information of people, and to evaluate and apply it in medical research. Here, particularly the genes of populations, who so far are living in isolation from other populations are highly coveted, and hence especially the genetic information of indigenous peoples. On the basis of their special genetic characteristics, the well known but also highly controversial *Human Genetic Diversity Project* promised to provide a clearer understanding of, for instance, heritable diseases and also of the prehistoric migrationary movements, natural selection and social as well as cultural developments.[42]

It is striking that biotechnological progress constantly produces new collective identities and immaterial assets, the legal situation of which is initially unclear. This relates in particular to genes which are not only rated as identity markers of individual persons but also constitutive of new genetic relationships by bringing together, for instance, patient groups with a common disposition to a certain disease:

> Such groups will have medical specialists, laboratories, narratives, traditions, and a heavy panoply of pastoral keepers to help them experience, share, intervene, and "understand" their fate. Fate it will be. It will carry with it no depth.[43]

According to Paul Rabinow's explanation we are seeing (non-meaningful) "biological" characteristics that unite people in new "social" groups with common interests: the genome constitutes "biosocial" communities of fate and in this way becomes the late modern substitute of the "soul".[44] Thus the controversies about the rights to genetic information become associated directly with human rights. This is especially apparent in the *Human Genetic Diversity Project*[45] as well as in

[41] Rojas, p. 36.

[42] Cavalli-Sforza et al., pp. 490 f.

[43] Rabinow, 'Biosociality', pp. 102 ff.

[44] Mauron, pp. 831 f.; Rabinow, 'Biosociality', pp. 91 ff.; id, 'Fragmentation', pp. 144 ff.

[45] Cavalli-Sforza et al., pp. 490 f.; Cann, pp. 443 ff.

comparable projects in other parts of the world. For example, the Icelandic human genome project caused a great stir when the genetic engineering company *deCODE genetics, Inc.* was founded.[46] Publicly raised accusations of "ecological colonization",[47] culminating in "racism"[48] and "vampirism"[49] show that such projects – despite good intentions to contribute to the general welfare of humanity – are obviously perceived as an existential threat by many people. They confront people with the notion completely contrary to the religious and holistic doctrines of indigenous peoples that their bodies could be transformed into isolated immaterial resources which are then publicly available as community assets. Hence they infringe particularly on traditional, religious, cultural and life-world concepts of physical and spiritual unavailability.

These examples show clearly that genetic resources are evidently of value not only for instrumental purposes. It seems that this is not solely a question of material, social and cultural welfare in the sense of a tangible benefit or efficiency gain. Apart from the economic and scientific interest in genetic information, a "genomic-metaphysical" thinking, widespread in Western societies, continues to exist which considers the genome to be the centrepiece of the individual identity of a single human-being and also of the collective identity of mankind as species. The identity-establishing power of the genome is based on its evolutionary interpretation as "common heritage". In consequence, the genetic resources must be preserved as such, in order that the evolutionary history can be retold in future.

But exactly this future is not predictable using the theory of evolution and has become still more doubtful with increasing biotechnological potential: nobody knows how the genetic constitution of humans and their animate world will develop, which specific needs people will have and which genetic resources will prove to be beneficial. It merely seems certain that new, often unexpected relationships and communities of fate will congregate in future.

3 Representatives of the Unforeseeable Future

Consequently the determination of whom or what has to be preserved in terms of biodiversity requirements for the future must be kept flexible. As a single constant, the original biological characteristic of a general genetic relationship remains, which links "biosocially" all past, present and future people to one another.

3.1 Junctions and Disjunctions of Law

Article 20 of the Constitution of the Federal Republic of Germany rests on this flexible identification of future communities of fate:

[46] Fortun, pp. 240 ff.; Pálsson and Rabinow, pp. 14 ff.

[47] Cunningham, p. 214.

[48] Alper and Beckwith, pp. 286 ff.

[49] Central Australian Aboriginal Congress, pp. 88 ff.

> Mindful also of its responsibility toward future generations, the state shall protect the natural foundations of life and animals by legislation and, in accordance with law and justice, by executive and judicial action, all within the framework of the constitutional order.[50]

This rule, formulated as a state aim, does not infer direct subjective legal rights, neither on people nor future generations, not to mention animals. Yet it binds state powers and plays a significant role especially in sovereign decisions when balancing competing interests.[51] Here, the problem of its indeterminacy becomes paramount when attempting to answer the question: which natural life bases should be conserved at what priority or what particular interests of future generations must be safeguarded? Although, according to the current legal situation, decisions regarding the conservation of natural resources are concerned with safeguarding *human* livelihoods, the target group becomes entirely indeterminate: "Whose future? Which generation?" – Who will be the future generations and what will they want?

The legal discourse evades these questions by tacitly assuming that future generations will in any event be people who fundamentally have the same interests and needs as the people of today.[52] Furthermore it is believed that the anthropocentric protection of the environment of today's people will prove advantageous for all other current and future, human and nonhuman beings. A further juridical fallacy is to grant rights to the present generation exclusively, simply because rights are created by people and are thus substantiated in the epistemic sense anthropocentrically. From this point of view, in principle only those can have rights who also issue legislation.[53]

However, none of these assumptions are compelling. For it would be no contradiction and would not be unusual for men as social, empathic beings,[54] from an anthropocentric or – more precisely – human-altruistic perspective, to protect nonhuman entities as well because of their intrinsic value according to ecocentric principles.[55] Such an approach would certainly entail an outcome troublesome not only to lawyers: apart from the legal claims of currently recognized legal subjects, humans and legal entities, possible and partly not yet identified interests of further actors not yet determined in detail would also have to be addressed.[56] These actors include nonhuman beings as well as – strictly speaking "not-yet-human" – future generations. Their expectations would probably also focus primarily on careful and sustainable management of natural resources by the present generation, with a particular interest in avoiding irreversible damage.[57]

---

[50] Translation published by the German Bundestag, available at: <https://www.btg-bestellservice.de/pdf/80201000.pdf> (visited 2. September 2010).

[51] v. Bubnoff, pp. 52 ff.

[52] Preuß, pp. 224 ff.

[53] v. Bubnoff, pp. 63 f.

[54] See e.g. Gallese, pp. 33 ff.

[55] Gruber, *Rechtsschutz*, pp. 179 f.; Krebs, pp. 19 ff.

[56] Latour, *Politics of Nature*, pp. 53 ff.; see also id., *Reassembling the Social*, pp. 155 ff.

[57] Preuß, p. 228.

This is indeed important for many people living today, and in a certain, holistic sense we, the "contemporary men" form a community of fate of all living beings with nonhuman and not-yet-human entities of future generations.[58] But the commonalities do not imply that we campaign effectively for the interest community of nonhuman and not-yet-human entities, too, out of our personal interest. Our altruism usually ends where our current, but in most cases short-range, interests would have to take second place in favour of long-term issues which will only take effect later. The phenomenon of an excessive valuation of short-term compared to long-term interests does not pertain solely to single economic decisions,[59] but also to "discounting" the interests of future generations.[60] Behavioural economists interpret this fact as a lack of self-discipline of people anxious for direct rewards on the one hand and a dependence of their intergenerational preference regimens on the way they frame the situation on the other.

This finding is scarcely surprising: of course, future human lives are, as a rule, valued less than present ones when surveyed in economic studies on the preference regimens of the value of human life. The dependence of such studies on the way the issue is framed indicates above all one thing: human lives cannot be offset against one another. If, nonetheless, an attempt is made, the result will have no prospect of general acceptance, as life does not follow the economic pattern of quantifiable and moreover highly selective benefit valuations which scarcely cope without externalization of costs – either at the expense of life-sustaining natural resources not appraisable on the market or at the expense of future generations.

Finally, not appraisable on the market is also the mentioned irreversible damage to future foundations of life. And exactly at this point it becomes apparent that it does not suffice to weigh risks exclusively according to the conventional cost-benefit equation, especially when dealing with innovative, partially still experimentally deployed technologies.[61] If irreparable damage is possible, this cannot be converted adequately into insurable cost risks. If such damage occurs, paying compensation to directly affected private persons and companies who sue for damage does not suffice. This is also the reason why the most recent oil disaster in the Gulf of Mexico cannot be "repaired", not even with a large sum of money.[62] From the outset, the indemnification currency "money" is the wrong one – and the compensation offered can never reach all aggrieved persons. For these not only include the actually affected persons and the business enterprises operated by them but also nonhuman life and living environments: animals, plants as well as habitats and their

[58] Rolston, pp. 13 ff.; Meyer-Abich, pp. 25 ff.

[59] Camerer, p. 10576 ("hyperbolic discounting").

[60] Revesz, pp. 987 ff.; see also Farber, pp. 289 ff. On this issue, see in the present volume Klaus Mathis, 'Discounting the Future?'.

[61] Busch, pp. 204 f.

[62] For this economistic error about the reparability of harm caused to the nature, see the statement of BP Chairman Carl-Henric Svanberg (16 June 2010): "We will look after the people affected, and we will repair the damage to this region, the environmental damage to this region and to the economy." Available at: <http://www.msnbc.msn.com/id/37739658/> (visited 2 September 2010).

life-support systems.[63] To do justice to all these within the framework of utilitarian concepts, the question regarding the affected stakeholders as legal subjects must be asked afresh.

3.2 Future Justice (3): Protecting Future Generations

Beyond the currently affected stakeholder groups, it is vital to take into account those whose impairment has not yet manifested and will possibly never be discernable with full certainty. In relation to biotechnological risks, safeguarding collective rights of traditional and rural communities to their knowledge and their contemporary culture is insufficient. Rather the issue is to leave scope for the development of new knowledge and future culture. This includes nature conservation sui-generis, abstracted from current needs. The preservation of biological diversity can thereby come into conflict not only with short-term interests of economic exploitation of natural resources but also with the current interests of local groups.

This also implies that safeguarding nonhuman and not-yet-human future generations can result in a reduction of "efficiency". However, what appears to be expensive and of no use from a common economic perspective, will in fact be a first step towards overcoming economistic reductionism. This perception of "efficiency" is highly selective to the extent that it is solely oriented towards current interests and ignores the externalization of costs, in particular those borne by future generations.[64]

In contrast, if we accept that we "present people" are linked in fate with uncertain "future people", then the efficiency calculations must also include their costs – as well as their benefits. However, these are unknown variables that possibly cannot be approximated, not even with contingency statements. If no one knows (and even cannot know) who the future generations will be, which communities of fate will evolve, which opposing parties will emerge and which interests they will develop, imponderables must be accepted: the calculated efficiency is applicable only under the proviso that it allows sufficient scope for possible, not anticipated and perhaps not at all predictable futures in order to permit future generations to make their new and own decisions.

However, the point cannot be to exceedingly restrict the freedom of presently living persons in favour of coming generations. In no way may the protection of future generations lead to the impossible requirement that every possible risk of irreversible damage must be ruled out. That would prevent almost every innovative technology. Rather, present freedom should merely be placed under the terms of future freedom.[65] Freedom conditioned in such a manner requires no more and no less than responsibility according to Kant's Moral Order:

---

[63] Gruber, *Rechtsschutz*, pp. 128 ff.; id., 'Rechte des Lebendigen', pp. 1546 ff.

[64] Preuß, pp. 228 ff.

[65] Preuß, p. 227.

> [A]ct only according to that maxim whereby you can at the same time will that it should become a universal law.[66]

With an eye towards future generations, this does not mean to dispense with all forms of action, the impacts of which will reach into the future. Instead, it is crucial that those who are affected, participate autonomously in decisions determining the future. If affected people are absent as is the case with future generations, others – as representatives – have to attend to their interests.

Such an advocacy of interests also presupposes that "someone" is represented. Doubtlessly, future generations have not yet reached the status of a "subject" – but at least in law the idea does not seem absurd to introduce new legal rights as legitimate claims of "legal subjects" with own interests. The law applies such constructions to correspond with its established terms to novel problems for which there are as yet no terms. If a legal term does not exist for future, nonhuman and not-yet-human communities of fate, then fictions[67] help to justify the exception from the rule that legal capacity only starts with birth, yet without undermining this rule. The law must reflect upon the (still) unthinkable,[68] in order to recognize the need for new concepts and develop these to address equitably newly arising issues. Even if future generations are (currently) no subjects, in jurisdiction – to be exact: in "jurisfiction"[69] – they can in principle be considered as legal subjects. However, legal fictionalization of such legal subjects cannot evade the question of what constitutes their identity. Future generations can be considered as having legal capacity only if they are characterized by certain attributes suggesting that they are independent stakeholders.

4 Future Generations as Community of Fate

There are four key aspects which can help classify what a "future generation" may signify and how future generations are constituted as communities of fate with a legal capacity.

4.1 Morally Unborn and Potential Life

Firstly, future generations appear as a parallel, collective form of unborn life forming the subject of bioethical debate in the field of reproductive medicine. In both of these cases – unborn children and embryos as well as future generations – it is a matter of *potential* humans and their *potential* futures. Their "moral birth" is occasionally

---

[66] Kant, p. 31.

[67] Esser, pp. 199 ff.

[68] For a famous example of thinking the "unthinkable", see Stone, pp. 450 ff.

[69] Esser, ibid.

brought forward in the debate to an earlier point in time, with the result that they are effectively treated as independent legal entities.

Considering that humans have emerged from the continuum of evolution of the entirety of life on Earth and continue to be part of it, this potential life is related to not-yet-human and even to nonhuman beings in some sense.[70] It is at least not illogical to endow those with own rights that could *potentially* evolve into personal identities.[71] However, as it is extremely rare to be able to determine whether an unborn or non-human animal effectively has the potential, such endowment must also apply to cases of doubt. These fundamentally must be treated in such a manner that the unborn are considered as legal persons: *in dubio pro persona*.[72] The requirements for recognition have to be broadly conceived so that the danger of excluding personal entities from the circle of legally protected persons does not arise. Otherwise it would have to be feared that, by rejecting the rights of those who are similar and related to us, our own personal status would no longer be permanently assured, but would have to be constantly substantiated anew.[73]

Therefore we can state: anyone who is close to us must receive own rights. "Closeness", however, does not presuppose solely a common ancestor from the continuum of evolution, nor just similarity with regard to certain biological characteristics. Rather a further, normative feature has to be added: ultimately, it is communicative integration into social relations that substantiates conditions of closeness and thus leads to the "moral birth" of entities with own rights.[74] Only through social communication will people be reconstructed as persons,[75] and unborn and future generations, too (and nonhumans, for that matter) have the prospect of being recognized as "legal subjects" only in this communicative form – the "Person Form".[76] Their recognition presupposes that the attribute is *ascribed* to them to be close to other persons. By communicative attribution people become our fellows just like unborn and future entities become our "neighbours in time".[77]

4.2 Collective "Quasi-Personal" Actors

Communication can produce social "closeness" and forge communities of fate not only among those present, but also among those absent: unborn and future generations. The normative consequence of such recognition of absentees would have to

[70] Gruber, *Rechtsschutz*, pp. 25 ff.

[71] Gruber, 'Kontinuum', pp. 131 ff.

[72] Hillgruber, pp. 975 f.

[73] Gruber, 'Kontinuum', pp. 138 f.; *Rechtsschutz*, pp. 115 ff.

[74] Gruber, *Rechtsschutz*, p. 133.

[75] Luhmann, *Social Systems*, pp. 210 ff.

[76] Luhmann, 'Form Person', pp. 166 ff.

[77] Preuß, p. 225.

be that the scope must remain at any time for them too, as potential stakeholders to achieve their development potential in future.

If this scope is now to be assured legally as well, then a second characteristic of future generations becomes clear: they occur as collective actors in legal communication which, while still indeterminate, evidently can be considered in principle legal entities. Whereas personification of collective actors is common in present law, for instance as communicative attribution (of legal status) to organizations, the potential reconstruction of future generations as legal entity presents particular difficulties. In general terms, the problem is that future generations are indeterminate not only with regard to their individual members, but also in respect to their collective identity as such, in spatial as well as temporal terms.

It is simply not predictable which specific interests future generations will have and which will be of such significance that they must already be enforceable under the present law. This non-predictability should not be taken to mean that these interests do not exist or can be ignored. Precisely this fallacy results from economistic reductionism by attempting to derive all relevant cost and benefit factors from present actors' interests and subsuming all decisions under the anthropocentric premise that natural life-support systems are identical with those of the people living today. The possibility that future interests could be different is systematically overlooked. No one can really know what it is like to be unborn. The ignorance of the reductionistic approach which likes to reduce problems into simple explanatory patterns, does indeed seem to be a constantly recurring problem.[78]

Not to know something should surely not be tantamount to ignoring it. With regard to the unknown interests of future generations, it follows that they could already be construed legally in the shape of potential interests, without necessarily needing to be spelled out in specific terms by a "stakeholder" taking a case to court in the present.

4.3 Variables for Contingent Associations

It follows that the third aspect of the constitution of future generations lies in the very indeterminacy in time, which provides the opportunity for the law to embrace potential fate communities of the future. Thus future generations may be considered as a variable for currently still contingent "associations" which will include human and nonhuman entities.[79] Such a variable provides scope for an expanded, non-anthropocentric interpretation of the concept of diversity. This must take into account the insight set out above, that today's people will no longer belong to future generations. Rather, the latter will include new, unexpected genetic relationships

[78]See Nagel, p. 437: "Any reductionist program has to be based on an analysis of what is to be reduced. If the analysis leaves something out, the problem will be falsely posed."
[79]Latour, *Politics of Nature*, pp. 57 ff.

and biotechnologically or "biosocially" based communities of fate and hence also include nonhumans.

From this, the requirement will surely result that truly sustainable conservation of biodiversity involves the conservation of the natural life-support systems of *all* living entities. Hence it will not suffice to take benefit for mankind alone as the benchmark of biodiversity valuation. For it is highly improbable that mankind's perceptions of usefulness and own wellbeing will remain stable over time, and just as improbable that they will match the interests of future generations.

Safeguarding biological diversity, and preserving natural bases for life, does not serve presently voiced interests alone, but is rather subject to communicatively generated norms in the social system. It is the expectations that have emerged in an evolutionary process[80] and have become established over time in society that must be stabilized not only for the continuance of social systems – but also for individual survival. It is these that must therefore be of universal human interest. Only in accordance with such expectations can individual "interests" be formulated at all and asserted in the shape of legal claims. In short, individual interests and claims do not exist by nature, but are created only in the communicative context of meaning of functioning societies. Here it is never only a matter of safeguarding the "own" and self-preservation, but beyond that of safeguarding the "neighbour" and sustaining social coexistence. And this is dependent on maintaining scope for a variety of actions and decisions in the future – achieved by assigning biological as well as cultural diversity an ecocentrically substantiated intrinsic value free of all utilitarian considerations.

In particular, one-sided, reductionistic approaches of partial social rationalities that allow definitions of value only according to their own standards, have to be counteracted. Claiming to be able to issue normative statements applicable to all spheres of the social system, ultimately prove to be destructive to society. They curb the scope of possibilities essential to the future by declaring their own rationality to be valid throughout the social system, their homogenization efforts effectively counteracting all kinds of diversity.

4.4 Useful Fiction for Future Justice

From this, the fourth character of future generations results. Their protection proves to be a useful fiction from an economic perspective as well: it enables the economic system to balance out on its own the deficits of unilateral economic approaches, to embrace alternative social rationalities and to transform these into economic communication – for instance in the form of "sustainable development" and the "precautionary principle".

---

[80] Luhmann, *Social Systems*, pp. 292 ff. and pp. 303 ff.

Whereas the precautionary principle is intended to confront the increasing risks of a contingent future, particularly in environmental and technology law,[81] a positive requirement is expressed with the notion of "sustainable development", as broadly outlined by the definition of the Brundtland Report:

> Sustainable development is development that meets the needs of the present without compromising the ability of future generations to meet their own needs.[82]

This definition continues to be indeterminate – however, that is as it should be. "Our common future" is equally ambiguous. Consequently, it is important from the constricted perspective of a societal sub-system, such as the economic system, to keep open the possible space for still unknown future communities of fate, their development potentials, potential interests and decisions.

Dealing with the unknown, coping with uncertainty, taking the imponderable into account – all this demands an impossible achievement: economy and law, in particular, must get an idea of something "other" that they effectively do not even grasp. "Our common fate with future generations" then appears to them as indeterminate, symbolic, even alien to the system. Therefore it is not surprising that, viewed from within the sphere of law, future-oriented conservation of biological and cultural diversity is considered less a legal or economic issue. It is rather conferred to other powers, notably to politics or even ethics.[83] But *perhaps* it is really a matter of law to adapt its own conceptualities to the requirements of its environment, to take present and future interests into account appropriately, in short: to do justice to the future.

5 Consequences for the Future of Law

It must indeed be questioned whether the body of global IP law, that is designed for worldwide expansion, is able to work against the described excesses of biopiracy. With its aspiration to enable worldwide enforcement and implementation of intellectual property rights, it laid the foundations for global exploitation of natural and cultural resources. The conflicting relationship between global intellectual property and biodiversity thus explains why international efforts to conserve and sustainably use biological diversity and assure an adequate sharing of benefits arising from utilization of genetic resources have had relatively little practical effect until now.

---

[81] Luhmann, *Risk*, pp. 29 ff.

[82] WCED, chapter 2: *Towards Sustainable Development*, sentence 1.

[83] Spranger, p. 92.

5.1 Biodiversity as an Issue of Future Law

For instance, the Convention on Biological Diversity (CBD)[84] drafted in 1992 as part of the Rio Declaration on Environment and Development can scarcely bear up against WTO law with such a target. Notably the target, formulated in Articles 15 and 16 of the CBD on Access and Benefit Sharing, is realizable only as far as the regulations of the more recent Agreement on Trade-Related Aspects of Intellectual Property Rights (TRIPS) allow: in particular Article 27(1) TRIPS which generally commits the nations party to the agreement to grant patent protection to all inventions, "provided that they are new, involve an inventive step and are capable of industrial application".[85]

Even if genetic resources are utilized without prior informed consent of the country of origin, and hence in infringement of Article 15(5) CBD, they must nevertheless be considered patentable inventions pursuant to the stipulations of TRIPS. The violation of CBD stipulations does not justify another interpretation offending against the possible meaning of the TRIPS agreements.[86] Attempts to call in question the "novelty" of an invention based on the acquisition of genetic resources or even view the original resource carriers as inventors are less than convincing. This circumstance does not change if, for instance, traditional knowledge is stored in databases in order to prevent it becoming the object of a "new" patent later.[87] The contexts in which traditional knowledge is produced cannot be maintained by administering their products in databases as isolated knowledge inventory alone – just as habitats cannot be saved by researching genetic resources in laboratories alone.

What must be added is a perspective orientated to the future: if the law embraces the fiction of future generations as communities of fate, it is able to reach beyond the borders of its own limited system of concepts and find answers to questions that commonly are considered to be in the political or moral realm. On behalf of future generations, law and economy need to establish their own boundaries by attaching an intrinsic value to the local "production centres" of traditional knowledge as well as biological diversity – a value which at least partially removes such centres from disposability.

5.2 Rights of Biodiversity

This "intrinsic value of biological diversity",[88] also mentioned in the preamble of the CBD, has a chance of being taken into account only if it is translated into

---

[84]See Art. 1 of the CBD, available at: <http://www.cbd.int/convention/articles.shtml?a=cbd-01> (visited 2 September 2010).

[85]See Agreement on Trade-Related Aspects of Intellectual Property Rights (TRIPS), available at: <http://www.wto.org/english/docs_e/legal_e/27-trips.pdf> (visited 2 September 2010).

[86]Kunczik, p. 101.

[87]Teubner and Fischer-Lescano, p. 19 and pp. 40 ff.

[88]See http://www.cbd.int/convention/articles.shtml?a=cbd-00 (visited 2 September 2010).

appropriate legal regulations and corresponding rights asserted by proper custodians. The countries of origin of genetic resources, designated by the CBD for this task, perform their function only partially, because their interests sometimes deviate considerably from those of the local communities. They often leave unused their legislative opportunities[89] to prevent the exploitation of genetic resources by adopting national conditions of access.[90] As a consequence it is important to involve those stakeholders that are more closely associated with the future development of the resources in question.

In the case of traditional knowledge, the local communities above all are especially interested in maintaining the continuity of productive knowledge contexts. Therefore, particular importance should be attached to their voice in decisions on access to natural and cultural resources. This could be achieved through the proposal by Gunther Teubner and Andreas Fischer-Lescano, who would like to assign "procedural", "communal-collective" rights to local communities as "epistemic groups" with the goal of making enforceable the necessary conditions for traditional knowledge production in the form of "discourse rights".[91]

In no way does this mean declaring the affected groups to be new holders of intellectual property rights in place of the "inventor". Instead of carrying on unchanged with the logic of "intellectual property", the affected groups should be appointed to act as guardians of indigenous "cultural rights" in order to represent, as advocates, the needs of future generations against an extension of IP law. From the perspective of IP law, the procedural participation of these representatives serves to limit its own validity claim in favour of foreign imperatives (under the reservation of *ordre public*). In doing so, its self-limitation must be sufficient to preserve indigenous cultures effectively.[92] This is the requirement placed on the law by future generations: to develop new terms and standards which allow self-restriction of global validity claims in law and economy to the benefit of other societal rationalities. Their representatives need to be heard as *altera pars*.

5.3 Rights of Future Justice

While it is relatively self-evident in the case of traditional knowledge to designate local communities as "correct" stakeholders, it is clearly more difficult to undertake the legal protection of natural resources under an advocacy aspect: the conservation of natural biocoenosis does not always match the current interests of local groups who may stand up for cultural diversity but not for biological diversity. Biodiversity does not serve solely economic utilization and social communication, but in addition carries a tangible weight for physical survival in the future. To do justice to the

---

[89]See Art. 15(1) CBD, available at: <http://www.cbd.int/convention/articles.shtml?a=cbd-15> (visited 2 September 2010).

[90]Kunczik, p. 100; Spranger, p. 92.

[91]Teubner and Fischer-Lescano, pp. 41 f.

[92]Teubner and Fischer-Lescano, pp. 40 f.

unique significance of biodiversity for the still contingent communities of fate of future generations, other suitable advocates therefore need to be found.

Present actors are almost overloaded with the requirement to enforce nature conservation abstracted from current needs. This involves altruism, relinquishment, and self-restriction in the interest of "intergenerational justice".[93] Accomplishing the task of future generations' representation of interests will be successful only if the problem is approached as a discursive collaborative project by local communities and global, governmental and also non-governmental organizations alike. Their disputes over the intrinsic value of natural resources, the ecocentric protection of non-human and not-yet-human life, not infrequently leading to a struggle for own necessities of life[94] and justified demands for the environment itself to receive legal standing,[95] at least open a window of opportunity to counteract the still advancing loss of biological diversity.[96] The concrete experience of existential threats may incur demands for justice for the future and may help to strike a balance between present and future needs – or in the words of Derrida:

> There is an avenir for justice and there is no justice except to the degree that some event is possible which, as event, exceeds calculation, rules, programs, anticipations and so forth.[97]
>
> ("Perhaps," one must always say perhaps for justice.[98])

Acknowledgement I would like to thank Gerda Häfner and Christopher Hay for their invaluable help. I am also grateful to Miriam Häfner, Klaus Mathis, and Gunther Teubner for their inspired comments on an earlier draft.

Bibliography

Alper, Joseph S. and Beckwith, Jon, 'Racism: A Central Problem for the Human Genome Diversity Project', in *Politics and the Life Sciences*, Vol. 18 (1999), pp. 285 ff.

Brand, Ulrich, 'Paradoxien der Biopolitik', in Karin Gabbert et al. (eds.), *Rohstoffboom mit Risiken*, Jahrbuch Lateinamerika 31 (Münster, 2002), pp. 127 ff.

Brand, Ulrich and Görg, Christoph, '"Nachhaltige Globalisierung"? *Sustainable Development* als Kitt des neoliberalen Scherbenhaufens', in Christoph Görg and Ulrich Brand (eds.), *Mythen globalen Umweltmanagements: "Rio + 10" und die Sackgassen 'nachhaltiger Entwicklung'* (Münster, 2002), pp. 12 ff.

Bubnoff, Daniela von, *Der Schutz der künftigen Generationen im deutschen Umweltrecht. Leitbilder, Grundsätze und Instrumente eines dauerhaften Umweltschutzes* (Berlin, 2001).

---

[93] Mathis, 'Future Generations', pp. 51 ff.

[94] See e.g. Complaint for Damages, *Native Village of Kivalina v Exxon Mobil Corp* (ND Cal, filed February 26, 2008), available at: <http://www.climatelaw.org/cases/country/us/kivalina/Kivalina%20Complaint.pdf> (visited 2 September 2010).

[95] Christopher D. Stone, pp. 457 ff.; Gruber, *Rechtsschutz*, pp. 160 ff.; id., 'Rechte des Lebendigen', pp. 1553 ff.

[96] Raven, pp. 29 ff.; see also Chen, pp. 50 ff.

[97] Derrida, p. 971.

[98] Derrida, ibid.

Busch, Lawrence, 'Commentary on Agricultural Biotechnology', in Charles McManis (ed.), *Biodiversity and the Law: Intellectual Property, Biotechnology and Traditional Knowledge*, (London and Sterling, VA, 2007), pp. 202 ff.

Camerer, Colin, 'Behavioral Economics: Reunifying Psychology and Economics', in *Proceedings of the National Academy of Sciences*, Vol. 96 (1999), pp. 10575 ff.

Cann, Howard M., 'Human genome diversity – Diversité génomique humaine', in *Comptes Rendus de l'Académie des Sciences – Series III – Sciences de la vie/Life Sciences*, Vol. 321 (1998), pp. 443 ff.

Cavalli-Sforza, Luigi L. et al., 'Call for a Worldwide Survey of Human Genetic Diversity: A Vanishing Opportunity for the Human Genome Project', in *Genomics*, Vol. 11 (1991), pp. 490 f.

Central Australian Aboriginal Congress, 'The Vampire Project: An Aboriginal Perspective on Genome Diversity Research', in *Search*, Vol. 25 (1994), pp. 88 ff.

Chen, Jim, 'Across the Apocalypse on Horseback: Biodiversity Loss and the Law', in Charles McManis (ed.), *Biodiversity and the Law: Intellectual Property, Biotechnology and Traditional Knowledge* (London and Sterling, VA, 2007), pp. 42 ff.

Cunningham, Hilary, 'Colonial Encounters in Postcolonial Contexts: Patenting Indigenous DNA and the Human Genome Diversity Project', in *Critique of Anthropology*, Vol. 18 (1998), pp. 205 ff.

Derrida, Jacques, 'Force de loi. Le "fondement mystique de l'autorité"' ('Force of Law : The "Mystical Foundation of Authority"'), in *The Cardozo Law Review*, Vol. 11 (1990), pp. 920 ff.

Egziabher, Tewolde Berhan Gebre, 'Bedrohte Ernährungssouveränität, internationales Recht und *Farmers' Rights* in Afrika', in Christoph Görg and Ulrich Brand (eds.), *Mythen globalen Umweltmanagements: "Rio + 10" und die Sackgassen 'nachhaltiger Entwicklung'* (Münster, 2002), pp. 154 ff.

Esser, Josef, *Wert und Bedeutung der Rechtsfiktionen. Kritisches zur Technik der Gesetzgebung und zur bisherigen Dogmatik des Privatrechts* (2nd edn., Frankfurt a.M., 1969)

Farber, Daniel A., 'From Here to Eternity: Environmental Law and Future Generations', in *University of Illinois Law Review* 2003, pp. 289 ff.

Fortun, Michael, *Promising Genomics: Iceland and deCODE Genetics in a World of Speculation* (Berkeley et al., 2008)

Gallese, Vittorio, 'The "Shared Manifold" Hypothesis: From Mirror Neurons to Empathy', in *Journal of Consciousness Studies*, Vol. 8 (2001), pp. 33 ff.

Gámez, Rodrigo, 'The Link Between Biodiversity and Sustainable Development: Lessons from INBio's Bioprospecting Programme in Costa Rica', in Charles McManis (ed.), *Biodiversity and the Law: Intellectual Property, Biotechnology and Traditional Knowledge* (London and Sterling, VA, 2007), pp. 77 ff.

Gruber, Malte-Christian, 'Vom Kontinuum der Herkunft ins Kontinuum der Zukunft. Zur Relevanz von Argumenten der Potentialität bei der Bestimmung des rechtlichen Status von Biofakten', in Nicole C. Karafyllis (ed.), *Biofakte. Versuch über den Menschen zwischen Artefakt und Lebewesen* (Paderborn, 2003), pp. 131 ff. (cited as: 'Kontinuum')

Gruber, Malte-Christian, *Rechtsschutz für nichtmenschliches Leben. Der moralische Status des Lebendigen und seine Implementierung in Tierschutz-, Naturschutz- und Umweltrecht* (Baden-Baden, 2006; cited as: *Rechtsschutz*)

Gruber, Malte-Christian, 'Die Rechte des Lebendigen: Wege zum Rechtsschutz nichtmenschlichen Lebens und natürlicher Lebensgesamtheiten', in *Aktuelle Juristische Praxis/Pratique Juridique Actuelle* (AJP/PJA), 12/2007, pp. 1546 ff. (cited as: 'Rechte des Lebendigen')

Hillgruber, Christian, 'Das Vor- und Nachleben von Rechtssubjekten. Über den Anfang und das Ende der Rechtsfähigkeit im öffentlichen Recht', in *Juristenzeitung* (JZ), 1997, pp. 975 ff.

Kant, Immanuel, *Groundwork of the Metaphysics of Morals*, translated and edited by Mary Gregor (Cambridge, UK, 1998)

Khush, Gurdev S., 'Biotechnology: Public-Private Partnerships and Intellectual Property Rights in the Context of Developing Countries', in Charles McManis (ed.), *Biodiversity and the Law: Intellectual Property, Biotechnology and Traditional Knowledge* (London and Sterling, VA, 2007), pp. 179 ff.

King, Steven R., 'Commentary on Biodiversity, Biotechnology and Traditional Knowledge Protection: A Private-sector Perspective', in Charles McManis (ed.), *Biodiversity and the Law: Intellectual Property, Biotechnology and Traditional Knowledge* (London and Sterling, VA, 2007), pp. 428 ff.

Kloppenburg, Jack and Rodríguez Cervantes, Silvia, 'Conservationist or Corsairs?', in *Seedling*, Vol. 9 (1992), pp. 12 ff., available at: <http://www.grain.org/seedling/?id=372> (visited June 11, 2010)

Krebs, Angelika, *Ethics of Nature: A Map* (Berlin et al., 1999)

Kunczik, Niclas, *Geistiges Eigentum an genetischen Informationen. Das Spannungsfeld zwischen geistigen Eigentumsrechten und Wissens- sowie Technologietransfer beim Schutz genetischer Informationen* (Baden-Baden, 2007)

Latour, Bruno, *Politics of Nature: How to Bring the Sciences into Democracy* (Cambridge, MA, 2004; cited as: *Politics of Nature*)

Latour, Bruno, *Reassembling the Social: An Introduction to Actor-Network-Theory* (Oxford and New York, 2005; cited as: *Reassembling the Social*)

Luhmann, Niklas, *Die Wirtschaft der Gesellschaft* (Frankfurt a.M., 1988; cited as: *Wirtschaft*)

Luhmann, Niklas, 'Risiko und Gefahr', in id. (ed.), *Soziologische Aufklärung 5: Konstruktivistische Perspektiven* (Opladen, 1990), pp. 131 ff.

Luhmann, Niklas, 'Die Form "Person"', in *Soziale Welt*, Vol. 42 (1991), pp. 166 ff. (cited as: 'Form Person')

Luhmann, Niklas, *Risk: A Sociological Theory* (Berlin and New York, 1993; cited as: *Risk*)

Luhmann, Niklas, *Social Systems* (Stanford, CA, 1995; cited as: *Social Systems*)

Luhmann, Niklas, 'Verständigung über Risiken und Gefahren', in id., *Die Moral der Gesellschaft*, ed., Detlef Horster (Frankfurt a.M., 2008), pp. 348 ff. (cited as: 'Verständigung')

Mathis, Klaus, 'Future Generations in John Rawls' Theory of Justice', in *Archives for Philosophy of Law and Social Philosophy* (ARSP), Vol. 95 (2009), pp. 49 ff. (cited as: 'Future Generations')

Mathis, Klaus, *Efficiency Instead of Justice? Searching for the Philosophical Foundations of the Economic Analysis of Law* (3rd edn., New York, 2009; cited as: *Efficiency*)

Mauron, Alex, 'Is the Genome the Secular Equivalent of the Soul?', in *Science*, Vol. 291 (2001), pp. 831 f.

Medaglia, Jorge Cabrera, 'The Role of the National Biodiversity Institute in the Use of Biodiversity for Sustainable Development – Forming Bioprospecting Partnerships', in Evanson C. Kamau and Gerd Winter (eds.), *Genetic Resources, Traditional Knowledge and the Law* (London and Sterling, VA, 2009), pp. 243 ff.

Meyer-Abich, Klaus Michael, *Praktische Naturphilosophie: Erinnerung an einen vergessenen Traum* (Munich, 1997).

Miller, Michael J., 'Biodiversity Policy Making in Costa Rica: Pursuing Indigenous and Peasant Rights', in *The Journal of Environment and Development*, Vol. 15 (2006), pp. 359 ff.

Nagel, Thomas, 'What Is It Like to Be a Bat?', in *The Philosophical Review*, Vol. 83 (1974), No. 4, pp. 435 ff.

Pálsson, Gísli and Rabinow, Paul, 'Iceland: The Case of a National Human Genome Project', in *Anthropology Today*, Vol. 15 (1999), No. 5, pp. 14 ff.

Preuß, Ulrich K., 'Die Zukunft: Müllhalde der Gegenwart?', in Bernd Guggenberger and Claus Offe (eds.), *An den Grenzen der Mehrheitsdemokratie. Politik und Soziologie der Mehrheitsregel* (Opladen, 1984), pp. 224 ff.

Quezada, Fernando, 'Status and Potential of Commercial Bioprospecting Activities in Latin America and the Caribbean', in *CEPAL – Serie medio ambiente y desarrollo* No. 132 (Santiago de Chile, 2007), available at: <http://www.eclac.org/publicaciones/xml/5/29455/LCL2742-P.pdf> (visited June 11, 2010)

Rabinow, Paul, 'Artificiality and Enlightenment: From Sociobiology to Biosociality', in id. (ed.), *Essays on the Anthropology of Reason* (Princeton, NJ, 1996), pp. 91 ff. (cited as: 'Biosociality')

Rabinow, Paul, 'Severing the Ties: Fragmentation and Dignity in Late Modernity', in id. (ed.), *Essays on the Anthropology of Reason* (Princeton, NJ, 1996), pp. 129 ff. (cited as: 'Fragmentation')

Raven, Peter, 'The Epic of Evolution and the Problem of Biodiversity Loss', in Charles McManis (ed.), *Biodiversity and the Law: Intellectual Property, Biotechnology and Traditional Knowledge* (London and Sterling, VA, 2007), pp. 27 ff.

Revesz, Richard L., 'Environmental Regulation, Cost-Benefit Analysis, and the Discounting of Human Lives', in *Columbia Law Review*, Vol. 99 (1999), pp. 941 ff.

Richardson, Leslie and Loomis, John, 'Total Economic Valuation of Endangered Species: A Summary and Comparison of United States and Rest of the World Estimates', in K. N. Ninan (ed.), *Conserving and Valuing Ecosystem Services and Biodiversity: Economic, Institutional, and social challenges* (London and Sterling, VA, 2008), pp. 25 ff.

Rodríguez Cervantes, Silvia, 'Biodiversitäts-Politik und lokale Gegenmacht – Das Beispiel Costa Rica', in Christoph Görg and Ulrich Brand (eds.), *Mythen globalen Umweltmanagements: "Rio + 10" und die Sackgassen 'nachhaltiger Entwicklung'* (Münster, 2002), pp. 137 ff.

Rojas, Juan Ramon, 'Plant Pirates in Costa Rica', in *The New UNESCO Courier* (November 2005), p. 36, available at <http://unesdoc.unesco.org/images/0014/001420/142021e.pdf> (visited June 11, 2010).

Rolston, Holmes, 'Value in Nature and the Nature of Value', in Robin Attfield and Andrew Belsey (eds.), *Philosophy and the Natural Environment* (Cambridge, UK, 1994), pp. 13 ff.

Rosenthal, Joshua, 'Politics, Culture and Governance in the Development of Prior Informed Consent and Negotiated Agreements with Indigenous Communities', in Charles McManis (ed.), *Biodiversity and the Law: Intellectual Property, Biotechnology and Traditional Knowledge* (London and Sterling, VA, 2007), pp. 373.

Shiva, Vandana, *Biopiracy: The Plunder of Nature and Knowledge* (Cambridge, MA, 1997)

Spranger, Tade Matthias, 'Indigene Völker, "Biopiraterie" und internationales Patentrecht', in *Gewerblicher Rechtsschutz und Urheberrecht*, Vol. 103 (2001), pp. 89 ff.

Stoll, Peter-Tobias, 'Access to GRs and Benefit Sharing – Underlying Concepts and the Idea of Justice', in Evanson C. Kamau and Gerd Winter (eds.), *Genetic Resources, Traditional Knowledge and the Law* (London and Sterling, VA, 2009), pp. 3 ff.

Stone, Christopher D., 'Should Trees Have Standing? – Toward Legal Rights for Natural Objects', in *Southern California Law Review*, Vol. 45 (1972), pp. 450 ff.

Stone, Glenn Davis, 'The Birth and Death of Traditional Knowledge: Paradoxical Effects of Biotechnology in India', in Charles McManis (ed.), *Biodiversity and the Law: Intellectual Property, Biotechnology and Traditional Knowledge* (London and Sterling, VA, 2007), pp. 207 ff.

Teubner, Gunther and Fischer-Lescano, Andreas, 'Cannibalizing Epistemes: Will Modern Law Protect Traditional Cultural Expressions?', in Christoph Beat Graber and Mira Burri-Nenova (eds.), *Intellectual Property and Traditional Cultural Expressions in a Digital Environment* (Cheltenham, 2008), pp. 17 ff.

Tribe, Laurence H., 'Ways Not to Think About Plastic Trees: New Foundations for Environmental Law', in *The Yale Law Journal*, Vol. 83 (1974), pp. 1315 ff.

United Nations Educational, Scientific and Cultural Organization (UNESCO), *International Year of Biodiversity 2010: Biodiversity is life – Biodiversity is our life*, available at: <http://unesdoc.unesco.org/images/0018/001866/186637e.pdf> (visited 29 July, 2010)

Wambugu, Florence, 'Biotechnology for Sustainable Agricultural Development in Africa: Opportunities and Challenges', in Charles McManis (ed.), *Biodiversity and the Law: Intellectual Property, Biotechnology and Traditional Knowledge* (London and Sterling, VA, 2007), pp. 174 ff.

World Commission on Environment and Development (WCED), Report of the World Commission on Environment and Development: Our Common Future (Oslo, 1987), available at <http://www.un-documents.net/wced-ocf.htm> (visited 29 July 2010)

Wullweber, Joscha, *Das grüne Gold der Gene. Globale Konflikte und Biopiraterie* (Münster, 2004).

Zent, Stanford and Zent, Egleé L., 'On Biocultural Diversity from a Venezuelan Perspective: Tracing the Interrelationships among Biodiversity, Culture Change and Legal Reforms', in Charles McManis (ed.), *Biodiversity and the Law: Intellectual Property, Biotechnology and Traditional Knowledge* (London and Sterling, VA, 2007), pp. 91 ff.

Cultural Heritage Preservation and Socio-Environmental Sustainability: Sustainable Development, Human Rights and Citizenship

Milena Petters Melo

The idea of Development stands today
like a ruin in the intellectual landscape.
Its shadow obscures our vision.
–Wolfgang Sachs

Not ideas, but material and ideal interests, directly govern men's conduct. Yet very frequently the "world images" that have been created by "ideas" have, like switchmen, determined the tracks along which action has been pushed by the dynamic of interest.
–Max Weber

Nowadays, by considering the principles of the 1992 Rio Declaration and the outcomes of the 2002 Johannesburg Summit, it is not difficult to understand an integrated approach to human rights and sustainable development: both are inextricably linked, complementary, multifaceted and mutually reinforcing, and they embrace civil, cultural, economic, political and social dimensions. Moreover, the advancement in international law leads to the observation that the rationale of cultural heritage protection (in its tangible and intangible aspects) was re-dimensioned to emphasize its importance to the enjoyment of human rights and the promotion of cultural diversity. From this perspective, the safeguard of cultural heritage is directly associated with sustainable development, as a *source* and a *resource* for democracy and socio-environmental sustainability, to the present and future generations.

This paper focuses on the protection of cultural and natural heritage to highlight the interplay between sustainable development, human rights, democracy and citizenship. It presents a brief analysis of the "development" and "cultural heritage"

M.P. Melo (✉)
Research Centre on European Institutions, University of Naples Suor Orsola Benincasa, Naples, Italy; Environmental Constitutional Law and Comparative Studies, University of Blumenau – FURB, Blumenau, Brazil; University of Curitiba – UNIBRASIL, Curitiba, Brazil; University of Salento, Lecce, Italy
e-mail: pettersmelo@libero.it

K. Mathis (ed.), *Efficiency, Sustainability, and Justice to Future Generations*, Law and Philosophy Library 98, DOI 10.1007/978-94-007-1869-2_7,

conceptual evolution in the context of international law, in particular regarding the United Nations system. The aim is to underline the connections between cultural heritage and socio-environmental sustainability, the importance of an integrated approach to sustainable development and human rights, and the need for empowerment via participative citizenship and synergies on both a global and local level.

1 Introduction

Since the second post-world-war two period, democracy in Europe has progressively evolved as a *constitutional democracy,* and the different nation-states have incorporated the principles of the welfare state, protection of human dignity and promotion of social rights.[1] At the same time, in the field of international relations the need for peace, international cooperation and protection of human rights has fostered the development of different international and regional instruments. The result of which has provided specific rights and even new areas of study and juridical action. Gradually, on both the national and international level, individual rights and civil liberties have been expanded into a set of fundamental rights which includes collective and diffuse rights for the protection of social, cultural and economic aspects of both private and public life.[2] The evolution of these *fundamental rights* of constitutional democracies – first in Europe and more recently in several countries in Latin America, Africa, Asia and Eastern Europe – and the development of the human rights system in international law characterize the *Age of Rights*, as described by Norberto Bobbio in his classic work: *L'età dei diritti*.

The second half of the last century was also referred to as the *Age of Development.* Like a towering lighthouse guiding sailors to safety, "development" once stood as the idea that oriented emerging nations during their journey through the post-war period. Democracies and dictatorships alike proclaimed development as the primary aspiration once colonial subordination had ended. The increased use of the term created a common ground for political actors on both sides of the spectrum, from right to left, and from the elite to the grassroots.[3]

In recent times, the growing complexity of modern society has brought to the centre of the international agenda the question of environmental degradation and the challenges of integrating different ways of life and material production. Consequently, sustainable development emerged as a global concern.

[1]This process of development of the fundamental rights on the democratic Constitutional States create the basis to organize that whole of rights and institutions that some scholars and the European Court of Justice refers as the *constitutional common heritage* or *common traditions of the European constitutionalism*. On this subject, see Onida and Häberle.

[2]Since the 1993 Vienna Convention the international community recognized the interdependence of the set of human rights, as *indivisible, inalienable and universal* rights that embrace individual, civil, political, cultural, economic, and social dimensions.

[3]Sachs, p. 4.

While globalization generates accelerated transformations, balancing traditions with the modern world's way of life becomes an unavoidable process. The intensification of intercultural relations produces a significant impact on both a local and global level, revealing consequences for democracy, protection of human rights and socio-environmental sustainability. In particular, increased immigration represents a substantial challenge for the inclusion and integration of immigrants of diverse backgrounds and the citizens they join.

In this context, the importance of cultural heritage, as a *source* and a *resource* for democracy and socio-environmental sustainability is vital. By the mid-twentieth century through the rise of human rights in international law, the rationale of cultural heritage protection and its perceived significance to humanity the advancement of the arts, sciences and knowledge, *was recalibrated to emphasize its importance to the enjoyment of human rights and the promotion of cultural diversity*.[4] Cultural heritage and its protection were no longer based on exclusivity, but rather on the intrinsic importance it had for people and individuals, with regards to their identity, their enjoyment of human rights and to the modalities and the quality of their own participation in the social system.

By focusing on the protection of cultural and natural heritage, this paper will highlight the interplay between sustainable development, human rights, democracy and citizenship. First, I will briefly analyse the "evolution" of the concept of development and cultural heritage protection in the context of international law, in particular regarding the United Nations system. Then, the links between cultural heritage and socio-environmental sustainability are outlined. Finally, the main focus will be on the importance of an integrated approach to sustainable development and human rights, supported by synergies on different levels.

2 The Age of Rights and Development

The second half of the last century, known as *the Age of Rights*, was also referred to as the *Age of Development*. In the name of development, the South has struggled to catch up with the North, experts descended on villages near and far, and millions of people were turned into wage earners and consumers. From this perspective, "development" entails much more than technical activity or a socio-economic endeavour – *it has become a perception that models reality, a myth that comforts societies, legitimates and justifies interventions, programs and projects,* and often *appears illusionary while provoking great passion.*[5]

Wolfgang Sachs notes that "the lighthouse of development" was erected after the Second World War. Following the breakdown of the European colonial powers, the United States launched the idea of development with a call to every nation to follow in their footsteps:

---

[4]Vrdoljak, p. 4.

[5]Sachs, p. 1.

> Since then, the relations between North and South have been cast in this mould: "Development" provided the fundamental frame of reference for that mixture of generosity, bribery and oppression, which has characterized the policies toward the South. For almost half a century, good neighbourliness on the planet was conceived in the light of "development".[6]

Gustavo Esteva observes that since 1949, when the U.S. President Harry Truman declared in his inaugural address of January 20, that the southern hemisphere was "underdeveloped",[7] "development" started its way into the universal lexicon, used not only in official declarations, but also in the language used by grassroots movements. The result was a new means of identification and *a new perception of both one's own self and also of "the other"*. According to Gustavo Esteva, "underdevelopment" began then, on 20 January, 1949:

> On that day, two billion people became underdeveloped. In a real sense, from that time on, they ceased being what they were, in all their diversity, and were transmogrified into an inverted mirror of other's reality: a mirror that belittles them and sends them off to the end of the queue, a mirror that defines their identity, which is really that of a heterogeneous and diverse majority, simply in the terms of a homogenizing and narrow minority.
>
> [. . .] Since then, development has connoted at least one thing: to escape from the undignified condition called Underdevelopment. For those who make up two-thirds of the world's population today, to think of development – of any kind of development – requires first the perception of themselves as underdeveloped, with the whole burden of connotations that this carries. Underdevelopment is a threat that has already been carried out; a life experience of subordination and of being led astray, of discrimination and subjugation.[8]

2.1 The Discovery of Underdevelopment and Cultural Homologation

From this perspective, "development" acquired a violent colonizing power, converting history into a programme with a necessary and inevitable destiny. Industrial production, which was only one method of social construct among many, in turn became the end game of a unilinear way of social evolution. This colonizing acceptance of development *gave global hegemony to a purely Western genealogy of history, robbing people from different cultures of the opportunity to define the forms of their social life.*[9]

---

[6]Sachs, p. 3.

[7]In his inaugural speech, Truman declared: "We must embark on a bold new program for making the benefits of our scientific advances and industrial progress available for the improvement and growth of underdeveloped areas. The old imperialism - exploitation for foreign profit - has no place in our plans. What we envisage is a program of development based on the concepts of democratic fair dealing".

[8]Esteva, p. 7.

[9]In anticipation of Trumanism, Esteva writes (p. 9), "In the third decade of the century, the association between development and colonialism, acquired a different meaning. When the British government transformed its Law of Development of the Colonies into the Law of Development

The word development gradually became part of the general vernacular, accumulating a variety of connotations. At the same time, the abundance of meanings ended up diluting its precise significance. In fact, very few words are as feeble, as fragile and as unable of giving substance and meaning to thought and behaviour.[10]

However, the fragility of the concept facilitates the ability of different actors to interject their own interpretations, interests and demands. Even though, on its own the term lacks a precise denotation, "development" is situated among popular and intellectual perception as an evocation of a net of meanings in which the person who uses it seems to be irremediably trapped: *since the word seems to involve a favourable change; gives the impression of a step from the simple to the complex, from the inferior to the superior, from worse to better.*[11]

But for two-thirds of the world's inhabitants, *the positive connotation of the word "development" is a constant reminder of exactly what they are not. It evokes a constant state that resides among the undesirable and undignified. To escape this condition, most of the world must become enslaved by someone else's dreams and experiences.*[12]

Therefore, it is important to underscore, as does the analysis by Gustavo Esteva, that underdevelopment is a comparative adjective whose base of support is the assumption, *very Western but unacceptable and in-demonstrable*, of the *homogeneity and linear evolution of the world. It displays a falsification of reality produced through dismembering the totality of interconnected processes that make up the world's reality and, in its place, it substitutes one of its fragments, isolated from the rest, as a general point of reference.*[13]

---

and Welfare of the Colonies in 1939, this reflected the profound economic and political mutation produced in less than a decade. To give the philosophy of the colonial protectorate a positive meaning, the British argued for the need to guarantee the natives' minimum levels of nutrition, health and education. A 'dual mandate' was sketched: the conqueror should be capable of economically developing the conquered region and at the same time accepting the responsibility of caring for the well being of the natives. After the identification of the level of civilization with the level of production, the dual mandate collapsed into one: development".

[10]Esteva, p. 8.

[11]Esteva, p. 10.

[12]Esteva, p. 11. On this subject, it is opportune to note that 'development' has always had a strong gender critique. There is a wealth of literature on gender and development; interesting analyses can be found in Shiva; Harcourt; and the selected papers of the SID (Society for International Development) website: http://www.sidint.net

[13]Esteva (pp. 11 ff.) adds that, "In Latin America, the Peace Corps, the Point Four Program, the War on Poverty, and the Alliance for Progress contributed to root the notion of underdevelopment into popular perception and to deepen the disability created by such perception. [...] But none of those campaigns are comparable to what was achieved, in the same sense, by Latin American dependency theorists and other leftist intellectuals dedicated to criticizing the development strategies that the North Americans successively put into fashion. By adopting in an uncritical manner the view to which they meant to be opposed, they gave a virulent character to the colonizing force of the metaphor. [...] The very discussion of the origin or current causes of underdevelopment illustrates to what extent it is admitted to be something real, concrete, quantifiable and identifiable: a phenomenon whose origin and modalities can be the subject of investigation. The word defines

2.2 Development

The use of the term *development* was from the outset inherently restrictive and often linked to economic growth. In many different situations *development* came to describe the personal income of citizens living in so-called underdeveloped areas.

Subsequently, the concept of *social development* was introduced in the United Nations' Reports on the World Social Situation. Although the concept was somewhat unclear in its exact meaning, it appears to be the intention that social and economical development should be addressed as separate and distinct realities. In 1962, the UN's Economic and Social Council (ECOSOC) recommended that both aspects of development should be integrated. During the same year, the UN sought to broaden usage of the term by adopting the idea of *growth plus change*, as outlined in its report titled *The UN Development Decade: Proposals for Action*. Thus began the *First UN Development Decade* (1960–1970), which set out to verify the social, cultural, economic, qualitative and quantitative dimensions of development by establishing that the key concept should be the improvement of people's quality of life. In subsequent years, the United Nations Research Institute for Social Development (UNRISD) was founded in 1963, and in 1966 ECOSOC adopted another resolution that recognized the interdependence of economic and social factors and the need for harmonizing economic and social planning.

By the *UN's Second Development Decade* (1970–1980), various efforts were undertaken to improve the concept of development by creating new branches of the term including: "participative development", "another development", "human-centred development", "integrated development", and a "basic needs approach".[14] During the same period, UNESCO promoted the concept of *endogenous development*. This concept contradicted conventional perception of development by rejecting the idea that industrial societies should be imitated by others. Instead, the idea of endogenous development, which emerged from a rigorous critique of the

---

a perception. This becomes, in turn, an object, a fact. No one seems to doubt that the concept does not allude to real phenomena."

[14]The idea of a human-centred development emerged in 1974, with the Declaration of Cocoyoc. This declaration was adopted by the participants in a UNEP-UNCTAD Symposium on the Pattern of Resource Use, Environment and Development, in Cocoyoc, Mexico, on October 1974. The declaration also emphasized the need for diversity and recognized that different countries should pursue different development strategies. Other goals of the declaration included self-reliance and the requirement of fundamental economic, social and political changes. Some of these ideas were expanded in the proposals of the Dag Hammarskjoid Foundation, which in 1975 advocated 'another development' and particularly a 'human-centred development'. Cf. Esteva (pp. 15 f.): "Following Johan Galtung's idea of development as 'the development of a people', many experts believed that citizens should have greater influence in the development process and that this should be, as UNESCO insisted, integrated development; in other words, it should be a total, multi-relational process that includes all aspects of the life of a collective, of its relations with the outside world and of its own consciousness". For further information, see UNRISD, *The Quest for a Unified Approach to Development*, 1980.

hypothesis of development "in stages" (Rostow), recommended tailoring the different approaches to each nation based on each nation's unique characteristics. This idea quickly gained popularity.[15]

In more recent times, *development* progressed both conceptually and politically to embrace the idea of *sustainable development (for) our common future*, as prescribed by the Brundtland Commission. With the relevant contribution of organizations and scholars from Africa, South America and India, a new interpretation of development emerged which was strictly connected with human rights, democracy and social and environmental sustainability. Stemming from different areas of knowledge, diverse institutions and heterogeneous social movements, this new literature converged on the importance of promoting human rights, first in the context of the definition and implementation of the Right to Development, and later in that of the definition and implementation of sustainable development.[16]

2.3 An Integrated Approach to Human Rights and Sustainable Development

The 1986 *UN Declaration on the Right to Development* affirmed that development is a human right. This statement was reinforced by the Vienna Declaration in 1993 at the UN World Conference on Human Rights; which proclaims that the *right to development is an inalienable human right and an integral part of fundamental human freedom.*

Subsequently, this perception was confirmed at the UN global conferences on population and development (Cairo), on women (Beijing), and at the World Summit on Social Development (Copenhagen).This shows that human rights have played a prominent role in international development cooperation since the early 1990s. United Nations global conferences have highlighted the crucial links between the three key goals of the UN Charter: peace, development and human rights. At the same time, increased importance has been given to linking development and human rights.

[15]Even if, as Esteva observes (pp. 15 f.), "little acknowledged, however, was the fact that this sensible consideration leads to a dead-end in the very theory and practice of development, and that it contains a contradiction in terms. If the impulse is truly endogenous, that is, if the initiatives really come out of the diverse cultures and their different systems of values, nothing would lead us to believe that development, no matter how it is defined, would necessarily happen, nor even an impulse leading in that direction. If properly followed, this concept leads to the dissolution of the very notion of development, after realizing the impossibility of imposing a single cultural model on the whole world, as a conference of UNESCO experts recognized in 1978".

[16]As Bonavides and Sousa Santos point out, the right to development was a contribution from Africa to the field of international human rights. First proposed by M'Baye in 1972, it was formally recognized by the African Human Rights Charter. See Bonavides, pp. 523 ff.; M'Baye, p. 220, and Sousa Santos, pp. 229 f.; regarding the complexity of development and its connections to rights, freedom and liberties, an important contribution can be found in Sen, *Development as freedom.*

During the 50th anniversary of the Universal Declaration of Human Rights, the United Nations Development Programme (UNDP) presented a policy document to promote wide discussion and broad awareness of the links between human rights and development, entitled *Integrating human rights with sustainable human development* (UNDP policy document, January 1998).

By reaffirming the *cross-cutting nature of human rights* and their *indivisible, inalienable and universal* characteristics, this policy document stressed that *the emphasis on one aspect of human rights cannot be used to detract from the promotion of any other aspect,* thus three levels of support for human rights were described. First, the United Nations Development Programme *works for the full realization of the right to development*, and its mandate for the eradication of poverty can be understood in this light. Second, UNDP *advocates the realization of human rights as part of sustainable human development*, an approach that places *people at the centre of all development activities*; while the central purpose was *to create an enabling environment in which all human beings lead secure and creative lives*, sustainable human development has been thus directed *towards the promotion of human dignity, and the realization of all human rights – economic, social, cultural, civil and political.* Third, UNDP devotes more of its programming activities to *good governance.* At the request of governments, UNDP implemented programmes aimed at *reforming legislatures, increasing the efficiency of the executive and strengthening the judiciary, promoting the quality of governance and the rule of law, transparency, accountability and decentralization.* In addition, UNDP governance programmes *strengthen participation in decision-making at the national and local levels*, working with national authorities and civil society organizations to promote civil and political rights.[17]

Today, by considering the principles of the 1992 Rio Declaration and the outcomes from the 2002 Johannesburg Summit, it is not difficult to understand an integrated approach to human rights and sustainable development by embracing an integral set of rights that protect natural and human resources and prioritize the fight against poverty, attune to the right of self-determination of the people, and promote and protect civil, cultural, economic, political and social rights.

Sustainable development has become a comprehensive concept entailing environmental protection, the rational use of natural resources, and the full realization of human rights and fundamental freedoms; it is the recognition that human rights and socio-environmental sustainability are inextricably linked, complementary and multidimensional, as well as interdependent and mutually reinforcing.

[17]Speth, pp. 4 f.

3 Natural and Cultural Heritage

The recognition of the link between nature and culture was one of the main innovative resolutions of the World Heritage Convention,[18] which was linked to the beginning of the environmental movement and the Stockholm Declaration on the Human Environment, both of which occurred in 1972. These two documents – a landmark for the protection of the cultural and natural heritage of humankind – opened a new chapter of international law and transnational social movements by recognizing new collective and diffuse rights, and by expanding their correlative legal ownership.

Another important innovative resolution of the World Heritage Convention was the creation of *world heritage* as a concern for all of humankind and its association with *intergenerational equity.*

Marking a new stage in the development of this area of international law, the traditional approach of "cultural property" (designed by the 1954 Hague Convention for the Protection of Cultural Property in the Event of Armed Conflict, for protecting individual interests) shifts to protecting the interests of society by safeguarding natural and cultural common goods. Law has evolved to determine the value to be protected by norms established by *present and future generations*; in other words, society as a whole, rather than the particular possessor of a certain object.

The significance of this change of approach, as several scholars have pointed out, is that "property" is a Western concept which does not necessarily address the needs of all peoples, since there are several examples of societies that do not recognize property as a social possibility; *rather than owning something, individuals belonging to these societies believe that they are owned by the environment around them.* The shift from "property" to "heritage", thus, *allows for other elements to be taken into consideration; in particular, cultural connections between objects and certain groups, and the internationalization of the issue, since the term "heritage" suggests a much broader concern, as it addresses the whole of mankind.*[19]

Through the 1972 World Heritage Convention, the term *cultural heritage* was created as an umbrella term to encompass three different types of heritage; namely sites, monuments and groups of buildings. In more recent times, while "cultural heritage" shifted from an aesthetics-based type of appreciation towards a more culturally relevant-based approach, *intangible heritage* has been gaining significance in

---

[18] *Convention concerning the Protection of the World Cultural and Natural Heritage*, UNESCO (1972). For further information about the World Heritage Convention's defining articles of cultural heritage (Art. 1) and natural heritage (Art. 2), see the analysis by Yusuf and Redgwell, in Francione and Lenzerini; see also the contributions of Lixinski in his helpful book review 'World Heritage and the Heritage of the World'.

[19] Cf. Lixinski, p. 378; moreover Lixinski notes that "another significant issue from the World Heritage Convention is that of the prevailing approach towards 'outstanding universal value' and the fact that it involves local communities in the process of identifying, presenting and nominating cultural and natural heritage sites. This is important because it reconciles local values and traditions with the universal significance of a particular site, which favours a more 'pluralistic' and 'diversity-oriented' approach in the choosing and managing of World Heritage sites".

the field, and gradually has been incorporated into the World Heritage System, culminating in the *Convention for the Safeguarding of the Intangible Cultural Heritage* (UNESCO, 2003). The widespread understanding today is that monuments cannot be regarded in isolation, but that they are complex and multidimensional manifestations of heritage, embodying both tangible and intangible elements. This expanded concept of heritage facilitates a holistic approach to both (material and intangible) fields, and reinforces the propensity to reconcile values and traditions with the universal significance of a particular site or intangible good, which favours a more *pluralistic* and *diversity-oriented* approach.[20]

In addition, the Convention for the Safeguarding of the Intangible Cultural Heritage recognized that intangible cultural heritage is a vital prerequisite for socio-environmental sustainability: *as a factor in bringing human beings closer together and ensuring exchange and understanding among them* (Preamble). The intangible cultural heritage refers to the practices, representations, expressions, knowledge, skills – as well as the instruments, objects, artefacts and cultural spaces associated therewith – that communities, groups and, in some cases, individuals recognize as part of their cultural heritage. This intangible cultural heritage, passed from generation to generation, is constantly recreated by communities and groups in response to their environment, their interaction with nature and their history (Art. 1).

The importance of safeguarding natural and cultural heritage has been integrated in to various international instruments, such as the *Declaration on the Responsibilities of the Present Generations Towards Future Generations* (UNESCO, 1997), which, following in the footsteps of the 1992 Rio Declaration on Environment and Development, asserted the necessity for establishing new, equitable and global links of partnership and intra-generational solidarity, and for promoting *intergenerational solidarity*[21] for the continuance of humankind and the preservation of life on Earth. This document states that:

---

[20] And it increases the complexity of the analyses, procedures and policies to protect the common heritage. As Lixinski observes (p. 380): "Even though these two fields (tangible and intangible) are in close interconnection, the international instruments related to them create different legal regimes. The inscription of a manifestation of heritage as either tangible or intangible heritage requires a choice as to the aim of protection, and the possibilities each system offers. This determination is to be done on a case-by-case basis."

[21] The responsibilities of the present generations towards future generations, stressed by recent documents such as the *Convention for the Safeguarding of the Intangible Cultural Heritage* (UNESCO, 2003) or the *Universal Declaration on Bioethics and Human Rights* (UNESCO 2005), have already been referred to in various instruments such as the *Convention for the Protection of the World Cultural and Natural Heritage* (UNESCO, 1972), the United Nations *Framework Convention on Climate Change and the Convention on Biological Diversity*, adopted in Rio de Janeiro on 5 June 1992, the *Rio Declaration on Environment and Development*, adopted by the United Nations Conference on Environment and Development on 14 June 1992, the *Vienna Declaration and Programme of Action*, adopted by the World Conference on Human Rights on 25 June 1993, and the *United Nations General Assembly resolutions relating to the protection of the global climate for present and future generations* adopted since 1990.

> The present generations should strive to ensure the maintenance and perpetuation of humankind with due respect for the dignity of the human person. Consequently, the nature and form of human life must not be undermined in any way whatsoever (Art. 3). The present generations have the responsibility to bequeath to future generations an Earth which will not one day be irreversibly damaged by human activity. Each generation inheriting the Earth temporarily should take care to use natural resources reasonably and ensure that life is not prejudiced by harmful modifications of the ecosystems and that scientific and technological progress in all fields does not harm life on Earth (Art. 4). With due respect for human rights and fundamental freedoms, the present generations should take care to preserve the cultural diversity of humankind. The present generations have the responsibility to identify, protect and safeguard the tangible and intangible cultural heritage and to transmit this common heritage to future generations (Art. 7).

Thus, the advancement of international law leads to the observation that the rationale of cultural heritage protection (in its tangible and intangible forms) and its perceived significance to humanity through the progress of the arts, sciences and knowledge, *was re-calibrated to emphasize its importance to the enjoyment of human rights and the promotion of cultural diversity.*[22] Cultural heritage and its protection are no longer based on exclusivity, but rather on the intrinsic importance they have for people and individuals, as well as for their identity, their enjoyment of human rights and the modalities and the quality of their own participation in the social system.[23]

---

[22]Vrdoljak, p. 4.

[23]For a more complete comprehension of the evolution of international law in protecting cultural rights and heritage, see: *Universal Declaration on Human Rights*, United Nations, 1948; *International Covenant on Civil and Political Rights*, United Nations, 1966; *International Covenant on Economic, Social and Cultural Rights*, United Nations, 1966; *Recommendation concerning the Safeguarding of Beauty and Character of Landscapes and Sites*, UNESCO, 1962; *Declaration of the Principles of International Cultural Co-operation*, UNESCO, 1966; *Convention concerning the Protection of the World Cultural and Natural Heritage,* UNESCO, 1972; *Recommendation concerning the Protection, at National Level, of the Cultural and Natural Heritage*, UNESCO, 1972; *Recommendation concerning Education for International Understanding, Co-operation and Peace and Education relating to Human Rights and Fundamental Freedoms*, UNESCO, 1974; *Recommendation on the Participation of the People at Large in Cultural Life and their Contribution to It*, UNESCO, 1976; *Declaration on Fundamental Principles concerning the Contribution of the Mass Media to Strengthening Peace and International Understanding, to the Promotion of Human Rights and to Countering Racialism, apartheid and incitement to war*, UNESCO, 1978; *Declaration on the Right to Development,* UN General Assembly resolution 41/128, 1986; *Recommendation on the Safeguarding of Traditional Culture and Folklore*, UNESCO, 1989; *Declaration on the Rights of Persons Belonging to National or Ethnic, Religious and Linguistic Minorities,* UN General Assembly resolution 47/135, 1992; *Vienna Declaration and Programme of Action*, UN World Conference on Human Rights, Vienna, 1993; *Draft United Nations Declaration on the Rights of Indigenous Peoples*, UN draft, 1994; *Declaration on the Responsibilities of the Present Generations Towards Future Generations*, UNESCO, 1997; *Universal Declaration on Cultural Diversity*, UNESCO, 2001; *Convention for the Safeguarding of the Intangible Cultural Heritage*, UNESCO, 2003; *Declaration concerning the Intentional Destruction of Cultural Heritage*, UNESCO, 2003; *Convention on the Protection and Promotion of the Diversity of Cultural Expressions*, UNESCO, 2005; *Universal Declaration on Bioethics and Human Rights*, UNESCO, 2005.

From this perspective, the safeguard of cultural heritage is directly associated with sustainable development, as a *source* and a *resource* for democracy and socio-environmental sustainability, for the present and future generations.

3.1 Cultural Heritage, Social Inclusion and Environment

In social interactions, people's identity is shaped from a *cultural repertory* that enables individuals to form bonds based on a sense of belonging to a specific group or population. As the bond is formed, a process of self-identification begins. Social inclusion is then achieved based on the identification of common needs, rights and interests, and is associated with the realization of structures and institutions that are capable of meeting people's expectations.[24]

The safeguarding of cultural heritage, in both its tangible (public buildings, urban spaces, landscapes) and intangible forms (traditions, verbal memories, sense of places) stimulates a sense of belonging to a place. A living, democratic preservation of public heritage can strengthen a person's sense of belonging, both for locally born citizens and foreign-born immigrants. This in turn can improve everyone's quality of life. On the other hand, people feel a need to preserve an environment to which they feel a sense of belonging (this can apply to a specific place, nation, region or the planet).

At the same time, in our increasingly diverse societies, it is essential to ensure harmonious interaction among people and groups with plural, varied and dynamic cultural identities as well as their willingness to live together. *Policies for the inclusion and participation of all citizens are guarantees of social cohesion, the vitality of civil society and peace* (Universal Declaration on Cultural Diversity, Art. 2).

Cultural heritage – in the form of social models, their transformations and their related material expressions – represents that which past generations pass along to future ones. It provides the cognisance tools that facilitate an analysis of the present, an understanding of the past and a design for the future. Therefore, its protection is fundamental if an effective form of sustainable development is to be designed. Moreover, it is widely thought that progress and development are welcomed only if citizens are provided with places, structures and institutions enabling them to understand, communicate and personally elaborate their own socio-cultural identity.

The growing importance of protecting the common heritage in the context of globalization processes is understandable, given that cultural heritages are resources for the protection of cultural diversity and sense of identity in the face of growing standardization. In addition to this, cultural heritages as a resource with which to develop dialogue, democratic debate and openness between cultures, can induce a renewed awareness of the cultural identity dimension in conflicts. In this realm, the importance of education gains relevance, and cultural heritage can be used as

[24]See Attaianese, Coppola, Duca, Melo and Ferreira Jr.

a factor for peace, in interpersonal and intercultural dialogue, and for promoting mutual understanding and conflict-prevention (Convention of Faro, EU 2005).

Regarding sustainable development, cultural heritages are also precious resources in the integration of the different dimensions of development: cultural, ecological, economic, social and political. Therefore, cultural heritage is valuable for its own sake and for the contribution it can make to other policies.

From a macro-prospective, the protection of cultural heritage and the promotion of cultural diversity help to increase the possibilities – and improve the possible forms – for *building democratic societies that are just, participatory, sustainable and peaceful, both locally and globally* (The Earth Charter, Principle 3).[25]

3.2 The Common Heritage of Humanity: The Key to Sustainable Development

The Universal Declaration on Cultural Diversity (UNESCO, 2001) highlights the interaction between cultural heritage, sustainable development, human rights and participative citizenship.

This document also expressly identifies cultural diversity as *the common heritage of humanity,* a vital prerequisite for sustainable development, and *the key to sustainable human development* (Art. 11). As *a source of exchange, innovation and creativity* (Art. 1), cultural diversity *widens the range of options open to everyone; it is one of the roots of development, understood not simply in terms of economic growth, but also as a means to achieve a more satisfactory intellectual, emotional, moral and spiritual existence* (Art. 3). The protection of cultural diversity is an ethical imperative, inseparable from the respect for human dignity. It implies a commitment to human rights and fundamental freedoms, in particular the rights of persons belonging to minorities and those of indigenous peoples (Art. 4). Cultural rights are an integral part of human rights, which are universal, indivisible and interdependent. [...] all persons have the right to participate in the cultural life of their choice and conduct their own cultural practices, subject to respect for human rights and fundamental freedoms (Art. 5).

The Declaration on Cultural Diversity also reaffirms the importance of public policy in international, national and local realms in partnership with the private sector and civil society (Art. 11). While recommending social synergy, the Declaration also provides tools for the support of a "broader citizenship", meaning political and administrative participation in pluralistic societies. This concept is reaffirmed in several international documents designed to protect natural and cultural heritage,

[25] *The Earth Charter* was created by the independent Earth Charter Commission, convened as a follow-up to the 1992 Earth Summit, in order to produce a global consensus statement of values and principles for a sustainable future. The document was developed over nearly a decade through an extensive process of international consultation, to which a lot of people contributed. The Charter has been formally endorsed by thousands of organizations, including UNESCO and the IUCN (World Conservation Union), available at www.EarthCharter.org.

including the Convention for the Safeguarding of the Intangible Cultural Heritage (2003), which states that each State Party *shall endeavour to ensure the widest possible participation of communities, groups and, where appropriate, individuals that create, maintain and transmit such heritage, and to involve them actively in its management* (Art. 15).

The interplay between culture, development and human rights is evident in many recent international documents, including the *Convention on the Protection and Promotion of the Diversity of Cultural Expressions* and the *Universal Declaration on Bioethics and Human Rights*, both adopted by UNESCO in 2005.

The Convention on the Protection and Promotion of the Diversity of Cultural Expressions affirms in its preamble that *cultural diversity* is *a defining characteristic of humanity*:

> Being aware that cultural diversity creates a rich and varied world, which increases the range of choices and nurtures human capacities and values, and therefore is a mainspring for sustainable development for communities, peoples and nations,
>
> Recalling that cultural diversity, flourishing within a framework of democracy, tolerance, social justice and mutual respect between peoples and cultures, is indispensable for peace and security at the local, national and international levels,
>
> Celebrating the importance of cultural diversity for the full realization of human rights and fundamental freedoms proclaimed in the Universal Declaration of Human Rights and other universally recognized instruments,
>
> Emphasizing the need to incorporate culture as a strategic element in national and international development policies, as well as in international development cooperation, taking into account also the United Nations Millennium Declaration (2000) with its special emphasis on poverty eradication,
>
> Taking into account that culture takes diverse forms across time and space and that this diversity is embodied in the uniqueness and plurality of the identities and cultural expressions of the peoples and societies making up humanity,
>
> Recognizing the importance of traditional knowledge as a source of intangible and material wealth, and in particular the knowledge systems of indigenous peoples, and its positive contribution to sustainable development, as well as the need for its adequate protection and promotion [. . .]

The Universal Declaration on Bioethics and Human Rights outlines international commitments to the protection of the environment, the biosphere and biodiversity, as well as respect for cultural diversity, pluralism, human rights and democratic principles. As summarized in this declaration:

> Due regard is to be given to the interconnection between human beings and other forms of life, to the importance of appropriate access and utilization of biological and genetic resources, to respect for traditional knowledge and to the role of human beings in the protection of the environment, the biosphere and biodiversity (Art. 17); also to the importance of cultural diversity and pluralism should be given due regard. However, such considerations are not to be invoked to infringe upon human dignity, human rights and fundamental freedoms (Art. 12).

By emphasizing the value of cultural heritage as a vital factor in sustainable development directly linked with human rights, these international documents prove that respect for diversity and identity is inherent in the concept of sustainability. Both natural and cultural heritage, understood as elements of a common good, justify the widest possible democratic participation and the exercise of cultural citizenship. On the other hand, they also highlight the inescapable responsibilities towards that heritage.

4 A Problem of Effectiveness or Efficiency?

However attractive these theoretical and legal developments can be (and are), the unbalanced implementation of international law in the field of sustainable development and the lack of a normative structure for international economic relations between developing and developed countries, frequently emerges as a cause for concern. Particularly bothersome is the recurring absence of a "human face" to the International Economic Order.[26] Coupled with this is the need for constructive dialogue between governments and civil society regarding the development of standards of political and social freedoms and a human rights impact assessment as an integral part of international community projects and programmes related to sustainable development. The International Law Association has often been outspoken on these various issues.[27]

Assessing the positive and negative aspects of globalization within the context of sustainable development has been a primary focus of the international community, and has, on more than a few occasions, led to controversy. As Susana Camargo Vieira observes:

> The debate's central issue is simple: although many developing countries are part and parcel of the world economic growth process, the number of people living in absolute poverty is increasing. In addition, dramatic events in different parts of the world – especially in some regions of Africa, Europe and Asia, where new conflicts erupted and old ones continued – seriously threaten the path to sustainable development and peace and continue emphasizing the interdependent nature of peace and security; of the economic and the social; of respect for human rights; and of environmental conservation.[28]

From this perspective, the report presented by the Committee on Legal Aspects of Sustainable Development to the 69th ILA Conference in 2000, known as the London Conference, posed (but did not answer) a question that still reverberates: *Is international law moving in the direction of more justice and equity in international relations and for better public international decision making for a greater number of people? Does it serve the cause of sustainable development*?

---

[26] As stressed by the *Committee on Legal Aspects of Sustainable Development Report* to the 67th ILA Conference.

[27] The International Law Association has committees concentrating on different problems of the observance and implementation of Human Rights. It also has committees concentrating in problems of definition and implementation of Sustainable Development: which has Human Rights as one of its three pillars. The result of these Committees' work can be found in bi-annual Reports and Resolutions, which are available in the ILA website: www.ila-hq.org.

[28] Vieira, p. 3.

In fact, despite being actively promoted as a green and democratic redevelopment, it is important to note that in its mainstream interpretation, sustainable development has often been conceived as a strategy for sustaining "development", not for supporting the flourishing and endurance of an infinitely diverse natural and social life.[29]

At the same time, the *trade-off* between human rights and economic development[30] has been strongly criticized by several scholars, governments, social movements and many other people from diverse sectors, who view the Age of Rights as *a time of increasing rights violations*.[31] While the implementation of social, cultural and economic fundamental rights has been stagnated by "economic limits", there is a common-held belief, that the status of individuals and the restriction of civil rights – especially for immigrants, even in western democracies – is no less dreadful.[32]

In the context of globalization – and its potential for excluding and marginalizing a considerable segment of the international community's population by limiting people's resources and concentrating economic and political power – human rights and sustainable development become the new frontiers of democracy.[33]

Recently, the impact of a global financial and economic crisis has been significant, and no corner of the world has escaped the consequences. One result has been the realization that too much confidence in the economic system's principles and rules could be detrimental.[34] The exaggerated prioritization of market-oriented efficiency produced a relaxing of the political field's hegemonic superiority. But with the crisis, concerns have resurfaced regarding the human face of the international order and the importance of reinforcing, on a nation-state level, welfare and a guarantee of rights. The emergence of these concerns is significant, even if the impact has so far only been subtle.

Given existing disparities between and within different countries, and since sustainable development depends on cooperation between countries and within them (as stressed in Agenda 21),[35] the need to build a human rights culture within

---

[29]See Esteva and Sachs.

[30]Regarding the trade-off between human rights and economic development, see Donnelly, p. 180, and Sousa Santos, pp. 224 ff.

[31]A considerable amount of work has been published on this issue. Some excellent examples include: Amirante, 'I diritti'; Held; Herrera Flores; and Zolo, *Cosmópolis*. For a prolific analyses of the human rights paradoxes on the Latin American context see Proner.

[32]See Melo, 'Movimentos migratórios'.

[33]Cf. Balibar.

[34]See Ikeda, pp. 4 ff.

[35]The 1992 United Nations Conference on Environment and Development, held in Rio de Janeiro, was a significant milestone that set a new agenda for sustainable development. On that occasion, the international community recognized that environmental protection and social and economic development are fundamental to sustainable development. To achieve this development, the global programme entitled Agenda 21 and the *Rio Declaration on Environment and Development* were adopted, as well as the 2002 *Johannesburg Declaration on Sustainable Development.*

international, national and local institutions is crucial. *The right to participate in the development process, including the participation of the beneficiaries of development projects, is also crucial, as is the participation in the creation of mechanisms to effectively exercise those rights.*[36]

5 Human Rights, Sustainable Development and Citizenship

As the UN High Commissioner for Human Rights, Mary Robinson, observed:

> The 1993 World Conference on Human Rights and the 1995 World Summit for Social Development highlighted the importance of an integrated approach to social advancement. Lasting progress depends on respect for human rights and effective participation of citizens in public affairs. Nevertheless, it is also well known that democracy and human rights will prove elusive without social justice and sustainable development. Poverty deprives millions of their fundamental rights. Societies, in turn, are deprived of these people's contributions. Achieving sustainable progress requires recognizing the interdependence between respect for human rights, sustainable development and democracy.[37]

Viewed from this perspective, we are all potential actors in the design of sustainable development strategies, as well as custodians of human rights. While violations of these rights are a menace to humankind, they are both violated and protected by human beings. Therefore, human rights are a litmus test for humankind in its humanity.[38]

A significant aspect of the Universal Declaration of Human Rights is that it is the responsibility of every individual and every function of society to promote respect for human rights, secure *universal recognition and observance*, and interact in a spirit of brotherhood. The Universal Declaration proclaims that *everyone has duties to the community in which alone the free and full development of his personality is possible* (Art. 29). These concerns have been reinforced by several United Nations documents, and are extremely important in the context of sustainable development. Social capital and creativity are fundamental factors for development.

So, to improve the chances of sustainable development for our common future, what then is the best way to guarantee the progressive realization of human rights?

In the field of international law, several strategies have emerged to answer this question. First, the "violations approach", whereby human rights are closely monitored and abuses publicized in order to hold states accountable for not upholding the law and implementing international human rights commitments. A second method, which often complements the first, emphasizes a comprehensive view of human

---

[36]As stressed by the Committee *on Legal Aspects of a New International Economic Order final Report*, presented to 65th ILA Conference.

[37]Robinson, p. 6.

[38]It's important to underline that the ambivalence of the term is more striking in the Italian *umanità* and the Portuguese *humanidade:* both refer to 'humankind' and 'humanity' simultaneously in a descriptive and active manner. As Resta observes, "to be part of the humankind do not guarantee to every human being that particular feeling of *humanity*".

rights by stressing both the protection and promotion of rights. A third method, which has proven even more productive, involves the enforcement of laws on both a national and local level by the public powers-that-be, through governmental policies and participative governance strictly linked to citizens' demands for and promotion of their own fundamental rights. A more recent approach depends on global governance and partnerships between government, civil society and the private sector.

The best way to work for the solidification of human rights and fundamental rights and to establish appropriate standards for socio-environmental sustainability, therefore, should be an integrated approach toward creating a synergy between these different methods. They are complementary and have equal value.

Currently, international documents emphasize the necessity of involving all members of society in a rationale of democratic governance in all matters connected with cultural heritage and sustainable development.

Thus, while enforcing the rule of law is crucial for the protection of human rights and ensuring sustainable development, so too is the adoption of measures which enable people to exercise their rights under the law. Further requirements would be the empowerment of individuals and communities toward an *inclusive development*,[39] and the strengthening of government institutions to promote people's well-being based on their active participation in the life of society and on the fair distribution of benefits resulting from it. The result will be a participative citizenship in more democratic and pluralistic societies.

From this perspective, *citizenship* is much more than an entitlement concerning the formal ties with a nation-state, or the possibility of participating in decision making by voting on alternating and discontinuous opportunities. Citizenship, in this sense, is the right to be part and to participate in a community – it is defined by the *political rights* that are the necessary conditions for achievement of all fundamental rights, and at the same time the consequence of them.[40] By considering the development of human rights in international law and the set of fundamental rights in the Democratic Welfare States' constitutions, a broader concept of citizenship should guarantee *all rights* to *all people*. More than a political right, citizenship could be understood as a "*universal right to politics*".[41]

Last but not least, *politics* cannot only be the *administration of the status quo*. Politics would be much more successful by restoring the interpretation of politics rooted in Classical Antiquity: the *search for a good life*.[42] This interpretation seems to be more adapted to human rights' internationalism and the gradual recognition

[39]See Friedmann, and Baratta (all the works referenced).

[40]Regarding broadening the concept of citizenship in the sense of embracing all human rights proclaimed by international law, and the fundamental rights affirmed by countries' constitutions, see: Zolo, *La cittadinanza;* Amirante, *Cittadinanza*; Melo, 'Cidadania'; and of the same author 'A concretização'. Regarding the historical construction of the 'citizenship' in Europe see Costa; for the current legal stratification of the concept in Europe, see Amirante, 'I diritti'.

[41]Cf. Balibar.

[42]Cf. Baratta, 'Direitos Humanos'.

of a new juridical subjectivity: the *humanity*, while the people of the planet start to recognize themselves as subjects with the same needs, rights, interests, common goods and shared challenges.

6 Final Remarks

Careful study of the international documents results in the understanding that cultural heritage and pluralism are at the centre of a new perception of sustainable development, and are directly linked with human rights and democratic governance. But this new perception needs improvement and must be applied to apposite strategies that solidify these principles and rights.

Currently, sustainable development should be integrated into all fields of policy in order to realize *the goals of environmental protection, development and respect for human rights on international, national and local levels.* The *objective of sustainable development involves a comprehensive and integrated approach to cultural, economic, social and political processes* which seek to protect the right of all human beings to an adequate living standard based on their active, free and meaningful participation in development, and in the fair distribution of benefits resulting there from. *The fulfilment of an international bill of human rights that is comprised of economic, social and cultural rights, as well as civil and political rights, and the peoples' rights, is central to the pursuance of sustainable development.*[43]

The protection of world heritage and the promotion of cultural diversity, in this context, are crucial. The common heritage of humanity is a vital requirement for sustainable development as a source of exchange, innovation and creativity. *Cultural diversity is as necessary for humankind as biodiversity is for nature.* This diversity is embodied in the *uniqueness and plurality of the identities of the groups and societies comprising humankind. In dissociable from a democratic framework, cultural pluralism is conducive to cultural exchange and to the flourishing of creative capacities that sustain public life* (Declaration on Cultural Diversity, UNESCO, 2001).

Since culture is one of the mainsprings of development, *the cultural aspects of development are as important as its economic aspects, which individuals and people have the fundamental right to participate in and enjoy.* Cultural diversity is a rich asset for individuals and societies. *The protection, promotion and maintenance of cultural diversity are an essential requirement for sustainable development for the benefit of present and future generations* (Convention on the Protection and Promotion of the Diversity of Cultural Expressions, UNESCO 2005 – Guiding Principles, Art. 2).

Thus, it is essential *to reaffirm the importance of the link between* human rights, *culture and development for all countries* – particularly for developing countries –

[43]Cf. *New Delhi Declaration.* Such concept, however, emerges from a combinative analysis of international documents influenced and inspired by Rio '92.

and to support actions undertaken nationally and internationally *to secure recognition of the true value of this link* (Convention on the Protection and Promotion of the Diversity of Cultural Expressions 2005, Art. 1).

A central argument throughout this paper is a feasible philosophical justification for intergenerational equity, based on the fact that the human rights and the concern for environmental sustainability are today the greatest challenges and the best promises of modern society's rule of law. If *the idea of development stands today like a ruin in the intellectual landscape*,[44] and if human rights are a *monument of Modernity* constructed by blood, hope and illusions,[45] then *their shadows can obscure our vision*. However, they can also inspire our creativity to restore this heritage by joining legal, social, cultural and economic resources to consolidate better (intellectual, institutional and material) structures to sustain our *common future* and the *search of a good life*.

Bibliography

Amirante, Carlo, 'I diritti umani tra dimensione normativa e dimensione giurisdizionale?', in Lorenzo D'avack (ed.), *Sviluppo dei diritti dell'uomo e protezione giuridica* (Napoli, 2003), pp. 27 ff. (cited as: 'I diritti')

Amirante, Carlo, *Cittadinanza (teoria generale)*, in *Enciclopedia Giuridica*, Vol. XII. Istituto della Enciclopedia Italiana (Roma, 2004), pp. 1 ff. (cited as: *Cittadinanza*)

Attaianse, Erminia, Coppola, Nunzia, Duca, Gabriela, Melo, Milena Petters and e Ferreira Jr., Silvio Pinto, 'Sustainability and Social Inclusion in Cities: A Proposal of a Methodology for Material and Intangible Heritage Preservation', in Sérgio Lira, Rogério Amoêda, Cristina Pinheiro, João Pinheiro and Fernando Oliveira (eds.), *Sharing Cultures* 2009 – *International Conference on Intangible Heritage* (Pico Island, 2009), pp. 3 ff.

Balibar, Etienne, *Le frontiere della democrazia* (trad. it. Andrea Catone) (Roma, 1993)

Baratta, Alessandro, 'Criança, democracia e liberdade no sistema e na dinâmica da Convenção das Nações Unidas sobre os direitos das crianças', conference presented on the *Congress "Direito e Modernidade"* (Florianópolis, 1996)

Baratta, Alessandro, 'Direitos Humanos e políticas públicas', conference presented on the *Conferência Estadual de Direitos Humanos do Rio Grande do Sul* (Porto Alegre, 1998; cited as: 'Direitos Humanos')

Baratta, Alessandro, 'Bisogni e diritti umani', conference presented in the course *La costruzione culturale dei Diritti Umani*, promoted by the Istituto Italiano per gli Studi Filosofici (Naples, 2001)

Bobbio, Norberto, *L'età dei diritti* (Torino, 1990)

Bonavides, Paulo, *Curso de Direito Constitucional* (4th edn., São Paulo, 1993)

Costa, Pietro, *Civitas, Storia della cittadinanza in Europa*, Vol. 4 (Roma-Bari, 2001)

Donnelly, Jack, *Universal Human Rights in Theory and Practice* (Ithaca, 1989)

Earth Charter Commission, *The Earth Charter*, available at ww.EarthCharter.org

Esteva, Gustavo, 'Development', in Wolfgang Sachs (ed.), *The Development Dictionary – A Guide to Knowledge as Power* (Johannesburg, 1993), pp. 6 ff.

EU, 2005 *Council of Europe Framework Convention on the Value of Cultural Heritage for Society*, adopted by the Committee of Ministers at the 941st meeting of the Ministers' Deputies on 13

[44] Sachs, p. 1.

[45] Sousa Santos, p. 209.

October 2005, and opened for signature on 27 October 2005 at Faro (cited as: *Convention of Faro*)
Francione, Francesco and Lenzerini, Federico, *The 1972 World Heritage Convention: A Commentary* (Oxford, 2008)
Friedmann, John, *Empowerment: The Politics of Alternative Development* (Cambridge, 1993)
Häberle, Peter, 'I diritti fondamentali nelle società pluraliste e la Costituzione del pluralismo', in Massimo Luciani (ed.), *La democrazia alla fine del secolo: diritti, eguaglianza, Nazione, Europa* (Roma-Bari, 1994), pp. 94 ff.
Häberle, Peter, *Lo Stato costituzionale* (trad. it. F. Politi e S. Rossi) (Roma, 2005)
Harcourt, Wendy (ed.), *Development Collection: Gender and Human Development,* Vol. 53 (2010)
Held, David, *La democracia y el orden global* (Barcelona, 1995)
Herrera Flores, Joaquim, *El vuelo de Ateneo* (Bilbao, 2000)
Ikeda, Daisaku, SGI *Peace Proposal* presented to United Nations Organization on January 2009
ILA *Committee on Legal Aspects of a New International Economic Order final Report*, presented to the 65th ILA Conference (Cairo, 1992), Part IV point 4.3
ILA *Committee on Legal Aspects of Sustainable Development Report*, presented to the 67th ILA Conference (Helsinki, 1996), pp. 278 ff.
ILA *Committee on Sustainable Development Report* to the 69th ILA Conference (London, 2000), pp. 656 ff.
Lixinski, Lucas, 'World Heritage and the Heritage of the World', Book Review of F. Francioni and F. Lenzerini (eds.), The 1972 World Heritage Convention: A Commentary, Oxford, Oxford University Press, 2008, in *European Journal of Legal Studies*, Spaces of Normativity, Vol. 2, n. 1, 2008, pp. 371 ff.
M'Baye, Keba, 'Droits de l'homme et pays en dévelopment', in *Humanité et droit international* (Paris, 1991)
Melo, Milena Petters, 'Cidadania subsídios teóricos para uma nova praxis', in R. Pereira e Silva (ed.), *Direitos humanos como educação para a justiça,* pp. 77 ff. (São Paulo, 1998; cited as: 'Cidadania')
Melo, Milena Petters, 'A concretização-efetividade dos direitos sociais, econômicos e culturais como elemento constitutivo fundamental para a cidadania no Brasil', in Revista IIDH, n. 34-35, *Instituto Interamericano de Derechos Humanos* (San José Costa Rica, 2002), pp. 211 ff. (cited as: 'A concretização')
Melo, Milena Petters, 'Movimentos migratórios na era da globalização: entre igualdade e diversidade, as novas fronteiras da democracia', in J. R. Reis and R. Leal (eds.), *Direitos Sociais e políticas Públicas: Desafios Contemporâneos,* Vol. 9, pp. 2635 ff. (Santa Cruz do Sul, 2009; cited as: 'Movimentos migratórios')
Onida, Valerio, *La Costituzione ieri e oggi* (Bologna, 2008)
Proner, Carol, *Os direitos humanos e seus paradoxos: análise do sistema americano de proteção* (Porto Alegre, 2002)
Resta, Eligio, *Il diritto fraterno* (Roma-Bari, 2003)
Robinson, Mary, Presentation of the UNDP policy document *Integrating Human Rights with Sustainable Human Development,* a UNDP policy document, January 1998, available at: http://www.undp.org/governance/docs/
Sachs, Wolfgang (ed.), *The Development Dictionary – A Guide to Knowledge as Power* (Johannesburg, 1993)
Sen, Amartya, *Desenvolvimento como liberdade (Development as Freedom)* (São Paulo, 1999)
Shiva, Vandana, 'Resources', in Sachs, Wolfgang (ed.), *The Development Dictionary – A Guide to Knowledge as Power* (Johannesburg, 1993), pp. 206 ff.
Sousa Santos, Boaventura de, *La globalización del derecho. Los nuevos caminos de la regulación y la emancipación* (Santafé de Bogotá, 1999)
Speth, James G., Presentation of the policy document *Integrating Human Rights with Sustainable Human Development,* a UNDP policy document, January 1998; available at: http://www.undp.org/governance/docs/

Truman, Harry S. Truman, 'Inaugural Address, January 20, 1949', in *Documents on American Foreign Relations* (Connecticut, 1967; cited by Esteva), p. 7
UN, *Universal Declaration on Human Rights* (1948)
UN, *International Covenant on Civil and Political Rights* (1966)
UN, *International Covenant on Economic, Social and Cultural Rights* (1966)
UN, *Declaration on the Right to Development – UN General Assembly resolution 41/128 of 4 December 1986* (1986)
UN, *United Nations General Assembly Resolutions Relating to the Protection of the Global Climate for Present and Future Generations* (1990)
UN, *Framework Convention on Climate Change and the Convention on Biological Diversity* (Rio de Janeiro, June 1992)
UN, *Rio Declaration on Environment and Development*, United Nations Conference on Environment and Development (Rio de Janeiro, June 1992)
UN, *Declaration on the Rights of Persons Belonging to National or Ethnic, Religious and Linguistic Minorities* – UN General Assembly resolution 47/135 of 18 December 1992 (1992)
UN, *Vienna Declaration and Programme of Action* – UN World Conference on Human Rights (Vienna, 1993)
UN, *Draft United Nations Declaration on the Rights of Indigenous Peoples* (1994)
UN, *World Summit for Social Development* (1995)
UN, *United Nations Millennium Declaration* (2000)
UN, *Johannesburg Declaration on Sustainable Development* (2002)
UN, General Assembly document A/57/329. *New Delhi Declaration on Principles of International Law Relating to Sustainable Development*, presented by the Governments of Bangladesh and The Netherlands to the United Nations General Assembly and adopted in April 6, 2002 (cited as: *New Delhi Declaration*)
UNDP, *Integrating Human Rights with Sustainable Human Development* – policy document (1998)
UNESCO, *Recommendation Concerning the Safeguarding of Beauty and Character of Landscapes and Sites* (1962)
UNESCO, *Declaration of the Principles of International Cultural Co-operation* (1966)
UNESCO, *Convention Concerning the Protection of the World Cultural and Natural Heritage* (1972)
UNESCO, *Recommendation Concerning the Protection, at National Level, of the Cultural and Natural Heritage* (1972)
UNESCO, *Recommendation Concerning Education for International Understanding, Co-operation and Peace and Education relating to Human Rights and Fundamental Freedoms* (1974)
UNESCO, *Recommendation on the Participation of the People at Large in Cultural Life and their Contribution to It* (1976)
UNESCO, *Declaration on Fundamental Principles Concerning the Contribution of the Mass Media to Strengthening Peace and International Understanding, to the Promotion of Human Rights and to Countering Racialism, Apartheid and Incitement to War* (1978)
UNESCO, *Recommendation on the Safeguarding of Traditional Culture and Folklore* (1989)
UNESCO, *Declaration on the Responsibilities of the Present Generations Towards Future Generations* (1997)
UNESCO, *International Congress on Cultural Rights and Human Development Universal Declaration on Cultural Diversity* (2001)
UNESCO, *Convention for the Safeguarding of the Intangible Cultural Heritage* (2003)
UNESCO, *Declaration Concerning the Intentional Destruction of Cultural Heritage* (2003)
UNESCO, *Universal Declaration on Bioethics and Human Rights* (2005)
UNESCO, *Convention on the Protection and Promotion of the Diversity of Cultural Expressions* (2005)
UNRISD (United Nations Research Institute for Social Development), *The Quest for a Unified Approach to Development* (Geneva, 1980)
Vieira, Susana Camargo, *Building a Human Rights Culture for Sustainable Development – The ILA contribution,* Paper presented to the 73rd Conference of the ILA (Rio de Janeiro, 2008)

Vrdoljak, Ana Filipa, 'Cultural Heritage in Human Rights and Humanitarian Law' in O. Ben-Naftali (ed.), *Human Rights and International Humanitarian Law* (Oxford, 2009), available at: http://works.bepress.com/ana_filipa_vrdoljak/15, 67 pp.

Zolo, Danilo, *La cittadinanza: appartenenza, identità, diritti* (Roma, 1994; cited as: *La cittadinanza*)

Zolo, Danilo, *Cosmópolis: perspectiva y riesgo de um gobierno mundia* (Barcelona, 1997; cited as: *Cosmópolis*)

Part III
Law, Economics and Sustainability

Discounting the Future?

Cost-Benefit Analysis and Sustainability

Klaus Mathis

To determine the value of public projects, cost-benefit analysis is an important instrument for assessing the profitability of an investment. Discounting is a practice that most economists take for granted almost unquestioningly. Philosophers, on the other hand, largely seem to agree that it is unacceptable to place a lower value on future benefits and costs by means of discounting.

This has given rise to a vigorous debate over whether discounting should be allowed at all, and if so, at what level the social discount rate should be set. It is clear, for example, that the pure time preference of economic actors cannot provide any convincing justification for discounting. Greater weight is attached to the argument of social opportunity costs, i.e. the fact that the capital invested in any project could equally be invested elsewhere at a particular rate of interest.

Finally, making the link with sustainable development, a further issue must be borne in mind: discounting rests on the implicit assumption that all goods are commensurable and substitutable. Since this implies that natural capital is also substitutable, the question of discounting opens up the debate on "weak versus strong sustainability": in other words, is it tenable and permissible to treat natural capital as interchangeable with other types of capital? The indications are that the debate over discounting is skirting around the real problem.

1 Introduction

When decisions are made about public projects, cost-benefit analysis is an important instrument for evaluating the corresponding investment. As part of this, the discounting of benefits and costs is a practice that most economists take for granted almost unquestioningly. Philosophers, on the other hand, despite Birnbacher's claim that they rarely agree on anything, do seem to agree that there is no conceivable

K. Mathis (✉)
University of Lucerne, Faculty of Law, Frohburgstrasse 3, P.O. Box 4466, CH-6002 Lucerne, Switzerland
e-mail: klaus.mathis@unilu.ch

K. Mathis (ed.), *Efficiency, Sustainability, and Justice to Future Generations*, Law and Philosophy Library 98, DOI 10.1007/978-94-007-1869-2_8,

justification for discounting.[1] This position is generally supported with the argument that placing a lower value on the interests of future generations is incompatible with the principle of impartiality.

However, many differences of opinion are based on misunderstandings and unclear definitions. In this connection, Hampicke aptly coined the term "discounting fog".[2] The aim of the following analysis will therefore be to disperse as much of this fog as possible. To this end, the paper begins by explaining how a cost-benefit analysis is conducted (Section 2). Next it discusses the reasons usually cited for discounting (Section 3). It then tries to shed light on the real crux of the matter and propose a solution for reconciling economic rationality with the ethical postulate of intergenerational justice (Section 4). Finally, the conclusion will summarize the results of the analysis (Section 5).

2 Cost-Benefit Analysis

Cost-benefit analysis is the most important economic instrument for assessing the economic usefulness of any kind of project that a society proposes to undertake, from construction projects to environmental investments or regulations. Essentially it involves two phases: *quantifying the expected costs and benefits* and then *evaluating the results* of the overall calculus against some particular decision-making criterion.

In the first phase, the anticipated annual costs and benefits are determined for the full duration of a project. Fundamentally these are calculated according to the *willingness-to-pay* principle. On the benefit side, this is measured according to the maximum amount that the gaining party would pay for the benefit gained. Costs are measured according to the minimum amount that the losing party would accept in compensation. Where goods are traded in competitive markets, it is also possible to use *market prices* as a simpler means of measuring costs and benefits. However, this gets more difficult if no markets exist for the goods in question, which is frequently the case in the environmental sphere. Although the costs and benefits are less easy to determine here, the definitive criterion is still the principle of willingness to pay.[3]

Before the benefits and costs are balanced up, interest on the annual payment streams must be added or deducted up to a particular point in time. Usually all payment streams are discounted to the present time (or the start of the project). In this case, future payment streams are divided by the factor $(1+r)^t$, where r represents the rate of interest and t the number of years over which the investment is discounted. The balance of all payment streams discounted to the present day is known as the *net present value* (NPV).

---

[1]Birnbacher, 'Verantwortung', p. 103.

[2]Hampicke, 'Diskontierungsnebel', p. 127.

[3]Lind, 'Equity', p. 381.

The second stage of cost-benefit analysis consists of applying the *evaluation criterion*, according to which an investment should only be entered into if the *net present value is positive*. In the case of several mutually exclusive alternatives, the project which achieves the highest net present value should be selected. This is especially true if budget constraints mean that not all investments can be realized at the same time.[4] One drawback of net present value is that projects of different size and duration are difficult to compare. In that case, it is also possible to refer to another metric, the *internal rate of return* (IRR). Instead of fixing a discount rate and then calculating the net present value, this is determined by setting the net present value to zero and calculating what return the investment yields. Projects with a higher internal rate of return should be given preference over those with a lower internal rate of return.

It must also be borne in mind that the cost-benefit analysis is oriented to the criterion of allocative efficiency. But this does not mean Pareto efficiency whereby a change is not allowed to leave anyone worse off, but rather *Kaldor-Hicks efficiency*, according to which the discounted benefits of a project, on balance, are greater than the discounted costs. It makes no fundamental difference who makes a profit and who loses out, as long as the gainers value their gains more highly than the losers value their losses. In other words, the gainers would have to be able to compensate the losers *hypothetically*; actual compensation is not required, however.[5]

The consequence of discounting is that when benefits and costs which only occur in the future are discounted to the present day, their value is lower than if they were incurred immediately. The higher the discount rate r and the larger the number of discounting periods t, the more this effect will be magnified. For example, a sum of \$1,000 in 10 years' time, discounted at a rate of 5% is worth $1{,}000/(1+0.05)^{10} = \$614$ today. By the same method, \$1,000 in a hundred years' time is worth $1{,}000/(1+0.05)^{100} = \$7.60$ today. The same sum in two hundred years' time equals just 58 cents when discounted to the present day.

What discounting appears to do is to favour preceding over succeeding generations. Tyler Cowen and Derek Parfit illustrate this argument with the following example:

> Imagine finding out that you, having just reached your twenty-first birthday, must soon die of cancer because one evening Cleopatra wanted an extra helping of dessert.[6]

In any event, discounting opens the door to harming the environment at the expense of future generations, since the corresponding future costs are weighted lower in the cost-benefit analysis than they would be if they were incurred today. From our present perspective, even the most apocalyptic disaster would make little or no difference to our balance sheet if it will only occur in the very distant future.[7]

---

[4]Lind, 'Equity', p. 381.

[5]See Mathis, *Efficiency*, pp. 38 ff.

[6]Cowen and Parfit, p. 145.

[7]Birnbacher, 'Diskontierung', p. 118.

3 Reasons for Discounting

In the literature a number of reasons for discounting are generally discussed, namely uncertainty about the future, pure time preference, increasing prosperity and social opportunity costs.[8] These reasons will now be discussed in more detail.

3.1 Uncertainty About the Future

One of the functions of discounting is to take account of prognostic uncertainty, i.e. to incorporate risk and uncertainty factors into the assessment of present actions with future relevance.[9] The working assumption is that present costs and benefits are more certain than future ones, and therefore preferable if both are otherwise equal.[10] Birnbacher considers that such risk discounting not only justified but necessary, and this function should be declared as such.[11] Even Rawls who takes a critical view of discounting considers it justifiable to take account of uncertainties or changing circumstances.[12]

O'Neill, however, doubts whether any known probability function for future events exists. Often, he argues, the problem is that we are not dealing with risks, for which the probability of the event is known, but with uncertainty, where this probability is unknown. Consequently, it usually makes no sense to undertake risk discounting.[13] That aside, he sees no reason to assume, as discounting does, that risk and uncertainty would correlate systematically with intertemporal distance. He points out that much of the damage that might occur in the distant future is perfectly predictable, whereas many of the benefits attributed to the immediate future are very uncertain.[14]

3.2 Increasing Prosperity

A further argument in favour of discounting rests on the assumption that future generations will be better off than we are. If we also accept the law of diminishing marginal utility, discounting would be justified because we derive more utility from a unit of money today than in the future.[15] Lind therefore considers discounting necessary for reasons of intergenerational justice:

---

[8]Likewise e.g. O'Neill, pp. 49 ff.

[9]Birnbacher, 'Verantwortung', p. 105.

[10]O'Neill et al., p. 58.

[11]Birnbacher, 'Verantwortung', p. 105.

[12]Rawls, *TJ*, § 45, p. 295.

[13]O'Neill, pp. 50 f.

[14]O'Neill et al., p. 58.

[15]O'Neill, pp. 51 f.

> Can we justify current generations sacrificing 2-3% of GWP [gross world product] to increase the wealth of future generations who even after deduction for the high damage scenario are 2-15 times richer than the present generation? The answer is clearly no on the basis of intergenerational equity, which must weigh in favour of the current generation.[16]

This is a problematic argument, however, because whether future generations will enjoy higher prosperity is questionable; the true outcome might be just the opposite.[17] Furthermore, the definitive criterion would not be utility as such, but prices, i.e. willingness to pay for utility-bringing goods. Since willingness to pay tends to be higher in a wealthier society, even taking account of diminishing marginal utility there is no reason for it necessarily to be lower in future than it is today.

3.3 Pure Time Preference

A further argument is based on individual time preference. According to Robert W. Hahn, this is the most important reason for discounting:

> The basic rationale for discounting is that consumers are not indifferent between consuming a dollar's worth of a good today and one dollar next year; discount rates are necessary to reflect this preference.[18]

Yet, of all arguments, pure time preference is the most fiercely criticized. This goes back as far as Henry Sidgwick, who in "The Methods of Ethics" (1874) emphatically defended the view that future utility and present utility should be weighted equally; intertemporal distance should not have any effect:

> [T]he mere difference of priority and posteriority in time is not a reasonable ground for having more regard to the consciousness of one moment than to that of another.[19]

Following in this tradition, John Rawls makes reference to Sidgwick and clearly states:

> Unlike the principle of equality, time preference has no intrinsic ethical appeal.[20]

Rawls considers a time preference irrational at the level of the individual, and unjust at the level of society:

> In the case of the individual, pure time preference is irrational: it means that he is not viewing all moments as equally parts of one life. In the case of society, pure time preference is unjust: it means (in the more common instance when the future is discounted) that the living take advantage of their position in time to favour their own interests.[21]

---

[16] Lind, 'Equity', p. 384.

[17] O'Neill, p. 52.

[18] Hahn, p. 1026.

[19] Sidgwick, p. 381.

[20] Rawls, *TJ*, § 45, p. 294.

[21] Rawls, *TJ*, § 45, p. 295.

Pure time preference can also be discussed in terms of individual impatience or short-sightedness (myopia). These appear to be universal qualities, as Eugen von Böhm-Bawerk rightly commented in his famous treatise "Capital and Interest" (1884/89):

> How many an Indian tribe, with careless greed, has sold the land of its fathers, the source of its maintenance, to the pale faces for a couple of casks of "*firewater*"! Unfortunately very much the same may be seen in our own highly civilised countries. The working man who drinks on Sunday the week's wage he gets on Saturday, and starves along with wife and child the next six days, is not far removed from the Indian. [...] Which of us has not been surprised to find that, under the pressure of momentary appetite, he was not able to refuse some favourite dish or cigar which the doctor had forbidden—knowing perfectly that he was doing an injury to his health [. . .]?[22]

Even though such behaviour may be rational in the short term, it often causes people to do themselves harm in the long term. Of course it would be more beneficial to spread consumption evenly over a longer time span instead of consuming everything immediately. The short-sighted person realizes this as well, which is why he always regrets his short-termist behaviour in retrospect – and he most of all, because at any given moment he would always like to be consuming a lot.[23] Myopic individuals harm themselves in two respects: in return for the option of choosing the path of consumption, which already reduces utility, they deprive themselves of a further share of their resource allocation by giving it away in the form of interest.[24] Myopic behaviour is therefore inconsistent, and in that sense, not rational.[25]

3.4 Social Opportunity Costs

What economists mean by opportunity cost of an action is the sacrificed utility of the next-best alternative use of the resource in question. An efficient allocation of capital takes place when it is utilized to achieve the highest return. Not just economists but, for example, philosophers such as Rawls therefore favour the use of a discount rate in order to allocate the limited resources available for investment.[26]

Because these resources are scarce, their use gives rise to opportunity costs. The opportunity costs of an investment are measured fundamentally in terms of the return on alternative possible investments. Or to put in another way: if there is a positive rate of interest r, then more goods can be bought with a unit of money today than with the same unit of money in one year's time; one money unit x can be invested at interest rate r and in one year it will be worth (1+r)x. The positive interest

---

[22] von Böhm-Bawerk, *Positive Theorie*, Vol. 1, pp. 332 f.; quotation after von Böhm-Bawerk, *Positive Theorie (trans.)*

[23] Hampicke, 'Diskontierungsnebel', p. 129.

[24] Hampicke, *Ökologische Ökonomie*, p. 258.

[25] Hampicke, 'Diskontierungsnebel', p. 130.

[26] Rawls, TJ, § 45, p. 295.

rate is a fact that can be observed in every market economy. The main explanation for its existence is the *productivity of capital.*[27]

Assuming a perfect competitive market – free of market-distorting taxes and imperfections of the capital market – then there is one sole market interest rate r at which economic actors can borrow and lend money.[28] The opportunity costs of a public investment, if the state has to incur debt in order to finance it, are fundamentally measured according to the *interest rate the state has to pay on government bonds*. If the state investment "crowds out" more profitable private investments, however, the definitive factor is this higher return and not the interest rate.[29] The opportunity cost approach also implies that returns yielded by the investment can *consistently be reinvested at the opportunity cost interest rate.* This presupposes that appropriate follow-on investments are available.

4 Dispersing the Fog

4.1 What Is Discounted?

The dispute between philosophers and economists can partly be defused by clarifying, to begin with, what is being discounted. Philosophers think mainly of *welfare*, i.e. utility in a utilitarian sense, whereas economists mean goods that are valued in terms of prices or *willingness to pay*.[30] In his study on discounting, Broome has already drawn attention to this important distinction:

> Economists typically discount the sorts of goods that are bought and sold in markets [. . .]. Philosophers are typically thinking of a more fundamental good, people's *well-being*.[31]

In fact there is largely agreement – even among economists – that discounting, and hence down-valuing, the welfare of future generations contradicts the principle of impartiality and cannot therefore be justified. It is a different matter when discounting is applied to monetary streams, i.e. goods, or the associated profits and losses. The instrument of cost-benefit analysis works with this kind of discounting.[32]

This important distinction will be illustrated with the example of the economic valuation of human life. To determine the "value of life", there are various possible

[27] Von Böhm-Bawerk, *Positive Theorie*, Vol. 1, pp. 338 ff.

[28] Lind, 'Equity', p. 381; in greater depth in Lind, 'Major Issues', pp. 24 ff.

[29] Mishan and Quah, p. 149.

[30] Cf. Weikard, 'Diskontrate', p. 158.

[31] Broome, pp. 128 f.

[32] Parfit, p. 480, therefore expressly states that his critique of discounting is not meant to be aimed at the discounting of monetary values.

methods[33] which seek to establish either how much money an individual is willing to spend on measures to reduce the risk of death (the willingness-to-pay approach), or how much compensation they require in order to take the risk of death (the willingness-to-accept approach).[34] The following two examples are intended to explain this approach:

- Willingness-to-pay approach: Let us assume that out of 10,000 car drivers, one will die as the result of a traffic accident, i.e. each driver faces a statistical death risk of 1:10,000. Let us further assume that by paying $500 every one of them could eliminate the risk of death, thus reducing a risk of 1:10,000 to 0. Supposing that all of them were prepared to pay this amount, it would add up to a total of $5 million, which corresponds to the "value of life" in this situation.[35] However, since our general experience of the world tells us that there will always be some residual level of risk in road traffic which cannot be eliminated, we will assume below that by fitting an airbag at a cost of $500, the motorists concerned can only reduce their own risk of death by half. Anyone who is willing to pay this amount values his or her life at no less than $500/(0.5* 1/10,000) = $10 million.
- Willingness-to-accept approach: Suppose that job X pays a salary of $100,000 and job Y a salary of $120,000. In job Y, however, the risk of a fatal occupational accident is 1% higher than in job X. Anyone who is willing to accept job Y at the stated salary must implicitly place a value of at least $20,000/0.01 = $2 million on his or her life.

Most philosophers reject an economic valuation of human life on principle, believing that the value of a human life is infinite and should not therefore be monetarized. Due to the universal scarcity of resources, however, when the state comes to assess the cost-efficiency of regulatory measures for accident prevention, it has no choice but to carry out a cost-benefit analysis, in the course of which it has to put a value on the lives potentially saved or lost as a result of the measures. The costs of the measures to be assessed are consequently weighed against the value of the human lives concerned.[36]

The debate between philosophers and economists about the permissibility of valuing a human life thus revolves around a purely terminological problem which can easily be resolved by talking about *valuing the reduction of life-threatening risks* rather than the value of life.[37] For the aim is simply to determine empirically the amount that people are willing to pay in order to reduce fatal risks, or what

---

[33]The most commonly used methods are the "revealed preferences methods" (labour market analyses) and "consumer market behaviour methods" (product market analyses). On the various methods, see e.g. Brannon, pp. 60 ff. See also in this volume Balz Hammer, 'Valuing the Invaluable?'.

[34]Merten, p. 179.

[35]Viscusi, 'Value of Life', p. 201.

[36]Brannon, p. 60.

[37]Likewise Posner and Sunstein, p. 557.

amount of compensation they would require in order to accept such risks. Richard A. Posner puts it well:

> So while valuing human lives sounds like an ethical or even a metaphysical undertaking, what actually is involved is determining the value that people place on avoiding small risks of death and the transformation into a 'value of life' is done simply to determine how large the award of damages in a lump sum must be in order to be equivalent.[38]

It would also be completely wrong to derive a normative statement from empirical data: that would be committing a naturalistic fallacy. It should also be pointed out that the aim is not to determine the value of a certain specific human life, but that of a *statistical life* of the people potentially affected by the measure in question. And this is why the expression "the value of a statistical life" is often used in the relevant literature.[39] The cost-benefit analysis is not therefore differentiating intertemporally between the value of human lives; *only monetary amounts are being discounted, not lives as such.*

4.2 Opportunity Cost Versus Social Discount Rate

As we have seen, the arguments against discounting are mostly wide of the mark once we recognize the difference between discounting welfare and discounting monetary values. But there still remains the question of the "right" discount rate.

Among economists it is undisputed that for private investments,opportunity costs are the definitive factor. What is disputed is whether decisions about public projects should likewise be based on the interest rate on the capital market, or on an artificially determined rate known as the "social discount rate". The concept of the social discount rate is based on the Ramsey formula[40]:

$$r_t = \delta + \eta(C_t)^* \dot{C}_t / C_t$$

Under this formula there are two components to the social discount rate r_t: the pure rate of time preference δ and the product of the marginal utility of consumption $\eta(C_t)$ and the growth rate of consumption $\dot{C}_t/C_t$.[41] This takes account of the effect of declining marginal utility with rising income.[42] Nordhaus refers to this as "growth discounting".[43]

---

[38]Posner, p. 197.

[39]See Viscusi, 'Value of Life', pp. 196 and 201.

[40]For a more detailed examination of this issue, see Dasgupta, p. 279.

[41]Lind, 'Equity', p. 386 ($\dot{C}_t$ denotes the derivative with respect to t).

[42]Neumayer, p. 30.

[43]Nordhaus, *Global Commons*, p. 123.

Some economists suggest setting the pure rate of time preference to zero and paying attention to growth discounting only.[44] Along the same lines, Samida and Weisbach also call for the long-term discount rate to be oriented to long-term economic growth.[45] The suggestion of a reduced discount rate is well meant, but in practice it leads to inconsistencies since a lower discount rate – or no discounting at all – would result in more investments, which would require a significantly higher saving rate.[46]

Furthermore, the use of a reduced discount rate carries the risk that unprofitable investments will be made. In this respect a public investment is no different from a private one. Hence, future generations could even be harmed by a lower discount rate.[47] The World Bank supports this point of view:

> [W]e feel that meeting the needs of future generations will only be possible if investible resources are channelled to projects and programs with the highest environmental, social, and economic rates of return. This is much less likely to happen if the discount rate is set significantly lower than the opportunity cost of capital.[48]

Arbitrary political setting of the social discount rate is problematic from an economic perspective, too, because it infringes the principle that economic parameters should reflect the valuations of market participants (the principle of consumer sovereignty).[49]

And it is by no means the case that if the projects are realized, they will necessarily burden future generations with costs alone. Projects that emerge positively from the cost-benefit analysis achieve a return which is higher than the discount rate, otherwise they would not show any net present value. Hence the cost-benefit analysis ensures that only profitable projects are realized. The returns yielded by these projects multiply social welfare, and society can pursue a path of growth, which in turn increases the stock of social capital. Future generations will categorically attain a higher level of prosperity than they would have done if the projects had not been realized.

4.3 The Real Problem

But this line of argument, that any damage that future generations are likely to suffer can be compensated by the increase in welfare, has a small catch: the cost-benefit analysis works on the tacit assumption of the *substitutability of natural capital* with

---

[44] For example, Cline, p. 4.

[45] Samida and Weisbach, p. 169.

[46] Neumayer, p. 35.

[47] See also Sunstein and Rowell, p. 174.

[48] Birdsall and Steer, p. 6.

[49] Cf. Mishan and Quah, p. 172.

other types of capital. In other words, it is based on the paradigm of *weak sustainability.*[50] It is in this paradigm, and not in discounting, that the real problem of cost-benefit analysis lies.

The concept of weak sustainability demands the preservation of the entire capital stock. Put simply, this means that living off capital assets is not allowed. The capital stock K can be subdivided into natural capital K_N, mechanical capital K_M and human capital K_H. Therefore: $K = K_N + K_M + K_H = const.$ Natural capital includes raw materials, energy, nature reserves and landscapes, and ecosystems generally. Mechanical capital consists of capital in the narrow sense, e.g. tools, machinery, buildings or transport infrastructure. Human capital refers to human knowledge, skills and abilities.[51]

The different forms of capital can, in principle, be freely substituted (empirical interpretation) – and may indeed be substituted (normative interpretation). It follows that natural capital may be consumed if mechanical and/or human capital are commensurately increased. The rents from the consumption of natural resources must therefore be reinvested (known as the Hartwick rule).[52] The concept of weak sustainability tends to be combined with optimism about technology: an assumption is made that technological progress will overcome resource constraints.[53] This viewpoint is predominant among traditional economists.

The proponents of ecological economics are less optimistic in this respect. They doubt the ability of the market to bring about adequate substitution and innovation effects by means of price adjustment mechanisms.[54] In any case, they consider natural capital to be fundamentally unsubstitutable: they argue that ecological systems are not simply production factors that can be replaced with others; they are the very foundation of all life and economic endeavour. On this rationale, a concept of *strong sustainability* is required.[55] Accordingly, what should be preserved is not just the entire capital stock K, but specifically also the *stock of natural capital*: $K_N = const.$[56] *A loss* of natural capital cannot be balanced out by a greater stock of mechanical and/or human capital.

A more sophisticated view of strong sustainability is that only certain parts of the natural capital must be physically preserved, namely those which are effectively unsubstitutable. This is known as *critical natural capital.* Use of non-renewable resources would be permissible if part of the rent is invested in renewable energies. Destruction of the Amazon rainforest, on the other hand, would be unacceptable and

---

[50] O'Neill, p. 58; Neumayer, pp. 37 ff.

[51] Weikard, *Wahlfreiheit*, pp. 54 f.

[52] Weikard, *Wahlfreiheit*, pp. 55 f.

[53] Neumayer, p. 23.

[54] Nutzinger, p. 71.

[55] Costanza and Daly, pp. 2 f.

[56] Weikard, *Wahlfreiheit*, p. 55.

the diversity of animal and plant species would need to be conserved. The general principle is to ensure that the functions of nature remain intact.[57]

The whole debate over discounting is therefore skirting around the real problem. The critical issue is not the quest for the "right" discount rate, but to ensure that projects meet the requirements of strong sustainability. O'Neill hits the nail on the head:

> It is not that discount rates should be zero, negative or positive, but that they are on the whole *irrelevant* to the discussion of the policy one should adopt to the future. There are good principles that govern our dealings with the future – that we minimize resource depletion, that we avoid irreversible changes, that we engage in sustainable economic activity and so on.[58]

So, the underlying assumption of cost-benefit analysis is that future generations can be compensated financially – for the consequences of global warming, for example, or the loss of natural habitats or animal species. But it should be self-evident that this makes no sense in relation to the resources that support life itself. O'Neill once again:

> What matters is not a nominal sum of wealth, but the goods, in particular the environmental goods, it can be exchanged for. What does it profit a man who gains a fortune and yet loses the world.[59]

Even if one believes this compensation to be possible, there would have to be some way of making sure it is transferred from generation to generation, since it only takes a single spoilsport generation to consume the transfer, thus squandering the compensation.[60]

Therefore non-discounting is totally inappropriate as a means of guaranteeing sustainable development. But equally, the idea of a "correct" social discount rate also turns out to be a chimera. The discounting fog only starts to lift once it has been recognized that cost-benefit analysis implies perfect substitution of all types of capital. What is needed, therefore, are additional restrictions such as the preservation of critical natural capital. Talbot Page:

> In the intergenerational context we use both majority and potential Pareto improvement as decision criteria. But we use both with constraints. As a constitutional matter we may require that certain things – for example, First Amendment rights – be a part of everyone's picture, regardless of the choice over some particular project. [. . .] In the intergenerational context the obvious constraints have to do with the quality of the resource base that is passed on from one generation to the next. This quality defines the basic opportunities that each generation has to work with, its means of survival, and, to a large extent, its well-being.[61]

But a lower discount rate, or even non-discounting, is neither a necessary nor a sufficient condition for the preservation of natural capital. Quite the opposite: if

---

[57]Neumayer, p. 25.

[58]O'Neill, p. 59.

[59]O'Neill, p. 59.

[60]See Mathis, 'Future Generations', p. 54.

[61]Page, p. 88.

it makes investments profitable which would have been unprofitable under market conditions, the danger of destroying natural capital is all the greater. Furthermore, it is by no means inevitable that normal discounting will only result in projects that are detrimental to future generations. In some circumstances it can also prevent environmentally harmful projects, since fewer projects will pass the cost-benefit analysis.[62]

4.4 A Two-Stage Decision-Making Procedure

As we have established, state investments like any others should be strictly oriented to the opportunity cost principle. The use of a reduced social discount rate may be a pragmatic way to favour projects from which future generations will benefit. Theoretically, however, this solution is not convincing.

The cost-benefit analysis should give us information on whether a project is profitable. If the discount rate is manipulated for non-economic reasons, the cost-benefit analysis can no longer answer even that question. And either way, it cannot help with the question of distributive justice. What purpose can it serve then? It tells us nothing useful, from an economic point of view or from an ethical point of view.

Therefore the question of discounting should not be mixed up with the requirement for sustainability and other ethical questions. Ethical arguments should be presented *as such* in a participatory and transparent decision-making process:

> A complementary approach is to incorporate noneconomic values through participatory decisionmaking after the strictly economic costs and benefits have been calculated and made public. But in neither case is it necessary or helpful to lower the discount rates.[63]

By openly declaring the ethical arguments alongside the cost-benefit analysis, it is possible to articulate the trade-offs between economic efficiency and ethical arguments and reveal the opportunity costs of the different alternatives.[64] The political decision-makers can then engage in an open discourse on the economic and ethical arguments. The use of a reduced discount rate, on the other hand, masks the value conflicts and shifts power in the direction of the technocrats who carry out the cost-benefit analysis and determine the "right" discount rate.[65] There is also a danger that the ethical arguments could become submerged in the overall calculation of the cost-benefit analysis.[66]

Equally, either discounting at a lower discount rate or non-discounting are not capable of eliminating the fundamental problem of cost-benefit analysis – the assumption that natural capital is substitutable. In addition to a cost-benefit analysis, therefore, all public projects should also be subject to a *separate sustainability*

---

[62] Cf. Viscusi, 'Rational Discounting', pp. 215 f.

[63] Birdsall and Steer, p. 7.

[64] On the concept of this trade-off, see Mathis, *Efficiency*, pp. 195 ff.

[65] Cf. Nordhaus, 'Public Policies', p. 158.

[66] Likewise Lind, 'Discount Rate', p. 457.

assessment. Investments that infringe upon critical natural capital are *a priori* not permissible, and consequently there is no need to carry out a cost-benefit analysis at all.[67] On the other hand, if investments do not infringe on critical natural capital, there is no objection to a cost-benefit analysis with discounting in accordance with the opportunity cost approach.

5 Conclusion

The dispute between philosophers and economists over the question of discounting can be defused, to begin with, by clarifying what is actually being discounted. Philosophers, by and large, base their assumptions on welfare, i.e. utility in a utilitarian sense, whereas economists mean goods that are valued in terms of prices or willingness to pay. An additional consideration of pivotal importance is that cost-benefit analysis implies the perfect substitutability of all types of capital (natural capital, mechanical capital, human capital). It is herein, and not in discounting, that the real problem lies. Consequently, restrictions are necessary in order to preserve critical natural capital. For this reason, public projects should not only be subject to cost-benefit analysis but also to a separate sustainability assessment.

Bibliography

Birdsall, Nancy and Steer, Andrew, 'Act Now on Global Warming – But Don't Cook the Books', in *Finance and Development*, Vol. 30 (1993), pp. 6 ff.

Birnbacher, Dieter, 'Intergenerationelle Verantwortung oder: Dürfen wir die Zukunft der Menschheit diskontieren?', in Reiner Kümmel and Jörg Klawitter (eds.), *Umweltschutz und Marktwirtschaft: aus der Sicht unterschiedlicher Disziplinen* (Würzburg, 1989), pp. 101 ff. (cited as: 'Verantwortung')

Birnbacher, Dieter, 'Lässt sich die Diskontierung der Zukunft rechtfertigen?', in Dieter Birnbacher and Gerd Brudermüller (eds.), *Zukunftsverantwortung und Generationensolidarität* (Würzburg, 2001), pp. 117 ff. (cited as: 'Diskontierung')

Brannon, Ike, 'What Is a Life Worth?', in *Regulation*, Vol. 27/4 (2004), pp. 60 ff.

Broome, John, 'Discounting the Future', in *Philosophy and Public Affairs*, Vol. 23 (1994), pp. 128 ff.

Cline, William R., 'Give Greenhouse Abatement a Fair Chance', in *Finance and Development*, Vol. 30 (1993), pp. 3 ff.

Costanza, Robert and Daly, Herman, 'Towards an Ecological Economics', in *Ecological Modelling*, Vol. 38 (1987), pp. 1 ff.

Cowen, Tyler and Parfit, Derek, 'Against the Social Discount Rate', in Peter Laslet and James Fishkin (eds.), *Justice between Age Groups and Generations* (New Haven and London, 1992), pp. 144 ff.

Dasgupta, Partha, 'Resource Depletion, Research and Development, and the Social Rate of Discount', in Robert C. Lind et al. (eds.), *Discounting for Time and Risk in Energy Policy* (Washington, 1982), pp. 273 ff.

---

[67] In a similar vein, see Kysar, p. 138.

Hahn, Robert W., 'The Economic Analysis of Regulation: A Response to the Critics', in *The University of Chicago Law Review*, Vol. 71 (2004), pp. 1021 ff.
Hampicke, Ulrich, 'Neoklassik und Zeitpräferenz – der Diskontierungsnebel', in Frank Beckenbach (ed.), *Die ökologische Herausforderung für die ökonomische Theorie* (Marburg, 1991), pp. 127 ff. (cited as: 'Diskontierungsnebel')
Hampicke, Ulrich, *Ökologische Ökonomie. Individuum und Natur in der Neoklassik – Natur in der ökonomischen Theorie* (part 4, Opladen, 1992; cited as: *Ökologische Ökonomie)*
Kysar, Douglas A., 'Discounting . . . on Stilts', in *The University of Chicago Law Review*, Vol. 74 (2007), pp. 119 ff.
Lind, Robert C., 'A Primer on the Major Issues Relating to the Discount Rate for Evaluating National Energy Options', in id. et al. (eds.), *Discounting for Time and Risk in Energy Policy* (Washington, 1982), pp. 21 ff. (cited as: 'Major Issues')
Lind, Robert C., 'Discount Rate & Social Benefit-Cost Analysis in Energy Policy Decisions', in id. et al. (eds.), *Discounting for Time and Risk in Energy Policy* (Washington, 1982), pp. 443 ff. (cited as: 'Discount Rate')
Lind, Robert C., 'Intergenerational Equity, Discounting, and the Role of Cost-Benefit Analysis in Evaluating Global Climate Policy', in *Energy Policy*, Vol. 23 (1995), pp. 379 ff. (cited as: 'Equity')
Mathis, Klaus, 'Future Generations in John Rawls' Theory of Justice', in *ARSP*, Vol. 95 (2009), pp. 49 ff. (cited as: 'Future Generations')
Mathis, Klaus, *Efficiency Instead of Justice? Searching for the Philosophical Foundations of the Economic Analysis of Law* (New York, 2009; cited as: *Efficiency)*
Merten, Carsten, *Die Bewertung menschlichen Lebens im Haftungsrecht* (Frankfurt a.M., 2007)
Mishan, E. J. and Quah, Euston, *Cost-Benefit Analysis* (5th edn., London and New York, 2007)
Neumayer, Eric, *Weak Versus Strong Sustainability. Exploring the Limits of Two Opposing Paradigms* (2nd edn., Cheltenham and Northampton, 2003)
Nordhaus, William D., 'Discounting and Public Policies That Affect the Distant Future', in Paul R. Portney and John P. Weyant (eds.), *Discounting and Intergenerational Equity* (Washington, 1999), pp. 145 ff. (cited as: 'Public Policies')
Nordhaus, William D., *Managing the Global Commons: The Economics of Climate Change* (Cambridge, 1994; cited as: *Global Commons)*
Nutzinger, Hans G., 'Langzeitverantwortung im Umweltstaat aus ökonomischer Sicht', in Carl Friedrich Gethmann et al. (eds.), *Langzeitverantwortung im Umweltstaat* (Bonn, 1993), pp. 42 ff.
O'Neill, John, *Ecology, Policy and Politics. Human Well-Being and the Natural World* (London and New York, 1993)
O'Neill, John et al., *Environmental Values* (London and New York, 2008)
Page, Talbot, 'Intergenerational Equity and the Social Rate of Discount', in V. Kerry Smith (ed.), *Environmental Resources and Applied Welfare Economics: Essays in Honor of John V. Krutilla* (Washington, 1988), pp. 71 ff.
Parfit, Derek, *Reasons and Persons* (Oxford, 1984)
Posner, Eric A. and Sunstein, Cass R., 'Dollars and Death', in *The University of Chicago Law Review*, Vol. 72 (2005), pp. 537 ff.
Posner, Richard A., *Economic Analysis of Law* (7th edn., New York, 2007)
Rawls, John, *A Theory of Justice* (Cambridge, 1971) (cited as: *TJ)*
Samida, Dexter and Weisbach, David A., 'Paretian Intergenerational Discounting', in *The University of Chicago Law Review*, Vol. 74 (2007), pp. 145 ff.
Sidgwick, Henry, *The Methods of Ethics* (7th edn. 1907, unedited reprint 1981, Indianapolis and Cambridge, 1981)
Sunstein, Cass R. and Rowell, Arden, 'On Discounting Regulatory Benefits: Risk, Money, and Intergenerational Equity', in *The University of Chicago Law Review*, Vol. 74 (2007), pp. 171 ff.
Viscusi, W. Kip, 'Rational Discounting for Regulatory Analysis', in *The University of Chicago Law Review*, Vol. 74 (2007), pp. 209 ff. (cited as: 'Rational Discounting')

Viscusi, W. Kip, 'The Value of Life in Legal Contexts: Survey and Critique', in *American Law and Economics Review*, Vol. 2 (2000), pp. 195 ff. (cited as: 'Value of Life')

von Böhm-Bawerk, Eugen, *Kapital und Kapitalzins. Zweite Abtheilung: Positive Theorie des Kapitales* (2 vols., 4th edn., Jena, 1921; cited as: *Positive Theorie*)

von Böhm-Bawerk, Eugen, *The Positive Theory of Capital*, trans. William A. Smart (London, 1891; cited as: *Positive Theorie (trans.)*)

Weikard, Hans-Peter, 'Soziale Diskontrate, intergenerationelle Gerechtigkeit und Wahlmöglichkeiten für zukünftige Generationen', in Hans G. Nutzinger (ed.), *Naturschutz – Ethik – Ökonomie: Theoretische Begründungen und praktische Konsequenzen* (Marburg, 1996), pp. 155 ff. (cited as: 'Diskontrate')

Weikard, Hans-Peter, *Wahlfreiheit für zukünftige Generationen. Neue Grundlagen für eine Ressourcenökonomik* (Marburg, 1999; cited as: *Wahlfreiheit*)

A Critical Review of "Efficiency Ethics": The Case of Climate Economics

Felix Ekardt

By examining the problem of climate change this paper develops (a) a substantial critique of the background assumptions that not only the formulation of economic theories but also, in parts, that of sociological/political theories is based upon, and (b) an approach to what should be understood by the term "ethics" with regard to the fifth Assessment Report 2014 of the Intergovernmental Panel on Climate Change (IPCC). However, ethics do not "supplement" "efficiency considerations" which up to now have dominated the practical results of the IPCC; they rather have to supersede them. It will be demonstrated that the supposed rationality behind the cost-benefit analysis used by economists and the IPCC – in correspondence to neo-classical economic theory – in order to more or less calculate mathematically the ideal climate policy, is only vaguely discernible, since both incorrect and incomplete normative and descriptive assumptions are incorporated into the calculation of what is supposed to be "efficient" climate policy. Accordingly, keywords are: predated and too optimistic climate data; problematic handling of prognosis uncertainty; missing important consequences of global warming such as wars over resources; not taking into account the limits of growth; improper quantification of what cannot be quantified; incorrect discounting of future events; and ethical and democratic deficiencies of "efficiency/preference theory" (to be clear, the problem lies within normative preference theory itself and not within the descriptive anthropology of the so called "homo economicus" which is often criticized in a rather misleading manner). The critique not only points at neoclassical environmental economics, Nicholas Stern, the IPPC, and, what is more, their "sceptical" critics, but also to some extent even at alternative economists. This paper also outlines an alternative to "efficiency thinking" which is not to be associated with "Rousseauian" or "Marxist" theories focusing on basic human needs or capabilities and Rawls' critique of utilitarianism. It therefore goes beyond the prevalent critique of the neo-classical approach to economics. A possibly more moderate but, from a methodological standpoint,

F. Ekardt (✉)
Baltic Sea Institute for Environmental Law, Research Group Sustainability and Climate Policy, Könneritzstrasse 41, D-04229 Leipzig
e-mail: felix.ekardt@uni-rostock.de

K. Mathis (ed.), *Efficiency, Sustainability, and Justice to Future Generations*, Law and Philosophy Library 98, DOI 10.1007/978-94-007-1869-2_9,

more coherent climate economics could be the objective that merges into a more general "climate social science" (*Klimasozialwissenschaft*) and a general balancing theory rather than focusing only on technicalities and natural sciences. Furthermore, the idea behind terms such as "ethics" and "theory of justice" that most social scientists have adopted will be corrected in the process of this review. These ideas are neither "vague", when it comes to their justification, nor do they solely correspond to the "democratic will of the majority". They are not even completely different from preference theories which have to be qualified as (less convincing) ethics themselves.

1 Introduction and Problem Specification

Science is the methodical and rational search for truth and/or justice – in the end, for its own sake. In the case where "facts" are to be found objectively we are talking about the "truth" and in the case where "norms" are to be justified objectively we are contemplating "justice". Climate (natural) science deals with facts; climate social sciences deal with facts and norms. As a part of climate social sciences, climate economics have played a more and more prominent role within the debate regarding both the "facts" and "norms" on which the appropriate climate policy should be based. Especially the economic ideas of Nicholas Stern and the Intergovernmental Panel on Climate Change (IPCC) have been very important and helpful for the climate debate during the last years. Even though there is more than one scientific approach to climate economics, all of them, as long as they can be assigned to the neoclassical approach to economics, are subject to more or less substantial and often simply over-looked criticism.[1] The critique presented here aims to establish comprehensive climate social sciences with a much broader view on the problem at hand, which is not restricted by economics.[2]

The subject of climate economics is the calculation of optimal climate policy paths; this also underlies the economic sections of the IPCC reports, where, so far economists are the only representatives of climate social science. For this purpose, the looming damage due to climate change and general advantages and disadvantages of climate policy steps, (for the most part) translated into monetary values, are placed in a relationship with each other. Looming damage due to climate damage, climate policy costs and climate policy benefits (translated into monetary values) are thus generally netted in economics in order to find the optimal path for climate protection.[3] The underlying principle applied is the idea of efficiency. This cost-benefit method from traditional welfare economics, however, has a fundamental problem.

---

[1] An alternative model to the neo-classical approach to economics would be ecological economics; see Daly; Rogall, pp. 157 ff. However, some of the following critique is also valid for this alternative concept.

[2] What is meant here is the underlying economics of climate protection and not economy as such.

[3] See, e.g., Posner, pp. 85 ff.; Nordhaus, p. 5.

"Exact data" in climate economics and in the IPCC may be convenient for many politicians and media representatives, and certainly appear to be so. Seemingly "exact data", however, disguise concealed information and assumptions about climate facts and normative goals. If these assumptions are wrong or questionable, the figures are worthless and ultimately a dishonest suggestion of objectivity. Even if "exact data" may scientifically – and even more so for political reasons and for the purpose of media coverage – appear attractive,[4] we shall see below that this economic approach is a disguised theory of justice, yet it is the dominant theory presented in the course of the climate debate. Unfortunately, the theory proves untenable in many important parts. Along the way, this paper thus criticizes the restriction of the term "science", which is commonly reduced among scientists and economists to (a) descriptive statements and especially (b) quantifiable information.

2 Realistic Climate Data, Economic Damages and Uncertainties?

First, we must briefly recapitulate the factual elements of the climate problem. This is not least necessary because there, already, the dominantly applied climate economic approaches are sometimes in themselves problematic, which can have consequences for justice matters, inter alia due to the possible underestimation of looming damage. Climate change is likely to challenge mankind with unprecedented problems. At its core concern, climate protection is rather simple, despite the scientific complexity of climate change[5]: the basic challenge is to emit far fewer greenhouse gases, i.e. (mainly) to consume a lot less oil, coal and gas. This requires strict greenhouse gas reduction targets, more energy efficiency, more renewable energy sources, which theoretically are largely free of greenhouse gas emissions, but perhaps also a certain amount of sufficiency. Thus a civilization model is at stake which, especially in the West in the last 200 years, has largely been funded on a high consumption of fossil fuels. In this model fossil fuels are omnipresent. Not just in gasoline and electricity, even in heat, in fertilizers, in almost every product, in plastics and transportation of goods. Issues such as high meat consumption, car trips and regular long distance holidays, overheated homes, and consumer electronics, to name a few, have therefore become part of the climate change discussion.

By the year 2100, assuming unchanged development, global warming is forecast to range between 3 and 6 degrees, possibly even more, especially if the emerging economies like China and India are increasingly successful in adopting the Western lifestyle. Without a much more rigorous approach to climate protection, the world is threatened with economic damage and threats to global peace as well as loss

[4]Critically, (however, only with regards to factual uncertainties) also Stehr and von Storch, pp. 19 ff.

[5]This formulation goes back to Hänggi, p. 7.

of life on a vast scale. At the heart of this lies a blatant global and intergenerational conflict[6]: despite the role often claimed by Europe and, especially, Germany as a "climate leader", data collected up to 2005 show that an average German person contributed to the production of about three times the greenhouse gas amount of a Chinese and about twenty times the amount of an African,[7] whereas southern countries will be affected relatively more by climate change in comparison to their northern neighbours.[8] The same applies to future generations: they are the victims of climate change without having caused it. Total global emissions since 1990 have increased by 40%. Even in Western countries, emissions have mainly been kept (only) constant, and even this has been done almost exclusively "crabwise" by accounting the collapse of Eastern European industry in 1990 and the (unintended) relocation of production to emerging countries as "domestic climate policy".

One often hears in political and scientific debates that global warming needs to be limited to no more than 2 degrees. Therefore it would be necessary to emit 60–80% less greenhouse gas in developed countries and 40–50% less worldwide by 2050 compared to 1990. However, the global climatological research, regularly consolidated in the IPCC, demands far more radical reductions to be able to avoid the possible catastrophic consequences with some certainty. The IPCC stated in its 2007 report that a 50–85% reduction of worldwide (!) greenhouse gas emissions from 2000 to 2050 was necessary if the intention was to accept global warming of no more than 2–2.4 (!) degrees. Yet it qualified even this (because of the feedback effects not covered) as probably still too cautious.[9] With a world population growing from 6.6 billion today to about 9 billion, this IPCC figure would require a per capita CO_2 emissions reduction of 4.6 tons per year (excluding deforestation) – and about 11 tons in Germany – to between 0.5 and 1 ton.[10] For industrialized countries, this would result in well over 90% emission reductions by 2050. Note that (1) this does not even take into account feedback effects, and (2) 2–2.4 degrees global warming may already imply substantial threats. In addition, (3) recent research related to the IPCC shows that the 2007 IPCC forecasts of climate change[11] will, in reality, be overtaken. Hence, from the perspective of climate science, the 2050 target for the Occident is basically a (nearly) zero-emission society, if one wants to avoid

---

[6]On the concept of sustainability (which means "more intergenerational and global justice") see Ekardt, *Theorie*; Ott and Döring.

[7]Cf. Baumert, Herzog and Pershing, p. 22.

[8]Böhringer and Welsch, p. 265. Nordhaus, p. 6, is rejecting any kind of consequences – in contrast to Stern, *Blueprint*, p. 13.

[9]On the following see IPCC 2007, p. 15, table SPM.5.

[10]See Hänggi, pp. 31 f., who calculates that according to the data of the IPCC in 2007 and in case world population rises to 9 billion by 2050 the per head out-put of CO_2-equivalents should be around 1.3–0.4 tons even without taking rebound-effects into account.

[11]Cf. the Copenhagen Synthesis from the beginning of 2009 (available at: http://climatecongress.ku.dk/pdf/synthesisreport); see also Hansen with regards to research conducted by NASA.

catastrophic damage.[12] Since human land-use emissions can never fall to zero, even negative emissions may be required for the energy sector, i.e. the recovery of greenhouse gases from the atmosphere.[13] All this is easily overlooked, since climate change is a delayed phenomenon and greenhouse gases sometimes remain in the atmosphere for centuries.

On the one hand, some damping feedback effects are already largely included in the climate models upon which climate forecasts are calculated. On the other hand, possible massively climate change reinforcing feedback effects are currently only inadequately covered. This concerns for instance: melting ice which can reflect less sunlight, increasing amounts of water vapour around the world due to increased temperature, the role of a change in cloud formation, the role of the oceans and the marine fauna, the release of greenhouse gas from thawing permafrost soil, and effects of climate change related changes in land use. Further calculation uncertainties exist in agriculture, particularly so in nitrous oxide and methane, and especially with respect to the global deforestation, which contributes to about 20% of climate change. Climate sceptics (who are not even climatologists) not only ignore that (1) the IPCC is rather cautious; they also exaggerate the degree of uncertainty in climate predictions and understate the predicted damage.[14] In addition, they regularly miss that (2) robust action is required solely due to the fact that fossil fuels are running out, even if, in the end, something less dramatic proves to be closer to the truth. Moreover, (3) climate sceptics in most cases neglect the precautionary principle: if one assumes that a dramatic risk to sensitive issues may be imminent and one knows that at the onset of the risk it will probably be too late for a remedy, it is advisable to act today. The latter, however, is a normative idea and assumes that there are normative issues that deserve protection. That this is indeed the case will be shown in Section 4.

The first problem of climate economics is that many climate economists appear relatively optimistic regarding the future development of climate change. Accordingly, they assume too little potential climate damage. Even the scientific foundation just described is not or not continuously represented throughout the preceding climate economics. At best, the IPCC 2007 data is used, which due to its methodology, reflects the state of knowledge from around 2004, and in that context,

[12]See the conclusion of EU's Council of Ministers (Environment) on 2nd of March, 2009 (available at: http://register.consilium.europa.eu/pdf/de/09/st07/st07128.de09.pdf) and the resolution of several managers of energy companies from April 2009, cited in TAZ on 10 April, 2009.

[13]It could, e.g., be feasible to combine bioenergy with CCS; cf. Ekardt, *Cool Down*, chapters 15 and 16.

[14]As an example for the following see only Lomborg. Climate sceptics ignore that some negative developments will occur with a delay of (at least) several decades as some greenhouse gases will stay in the atmosphere for a long time. Furthermore, because of the physical limits to growth the world will probably not for ever become richer and, therefore, we cannot assume that potential climate damages will simply be compensated by growing wealth. And, climate protection policy costs (in parts only alleged costs) are not better spend on the fight against Aids or malaria; we should do both, not only because climate change threatens to become the worst catastrophe developing countries have ever faced.

moreover, a more lenient scenario is often used. Even Nicholas Stern, who is considered perhaps the most influential climate economist and, in this case, often cited as exemplary and who likely exceeds a number of other climate economists in many ways, still refers in the summer of 2009 to a global reduction of only 50% by 2050, and does not even seem to have accounted for the Copenhagen Synthesis by IPCC members (2009). On the other hand, the Stern Review of 2006 points out that those figures are likely to be rather low. Thus, problematic factual assumptions become the basis for climate-economic calculations, which tend to underestimate the potential climate damages. This is all the more true if, in line with the views of the many Stern-critics such as William Nordhaus, the Stern assumptions are even rejected as exaggerated.[15]

Therefore, it is often overlooked that climate change – leading to crop failures, natural disasters, floods, water shortages, food shortages, areas and whole countries becoming uninhabitable, as well as vast migration flows – would be many times more expensive than the costs associated with effective action on climate change. Although the Stern Report in 2007 highlighted this realization, despite major resistance in the economic world,[16] recent calculations suggest it now actually proves to be too cautious.[17] Stern, on the other hand, criticizes that many economists do not adequately see the economic benefits of climate policy, that greenhouse gas limits, more efficiency, more renewable energy, and more sufficiency will in fact secure a permanent supply of electricity, a long-term supply of heat and fuel at acceptable prices, given the scarcity of fossil resources and the instability of some supplier states,[18] and even short-term savings in energy costs (such as thermal insulation) and opportunities for new jobs and markets through new technologies.[19] Beyond the question of current climate data, however, another major omission is present in the economic factual material, in the Stern Report, the IPCC as well as others: the maybe cynical-sounding, but perhaps most monetarily quantifiable cost does not seem to be considered – the cost of possible military conflicts over oil, water and other resources. If calculations are still rather too cautious, then this also documents how problematic even in purely economic terms the current political debate about "less climate protection because of the financial crisis" truly is.[20]

These suggestions for an "update" of climate economic calculations do not go so far as to reject a climate economic approach altogether but could be considered within an economic framework. A structurally unsolvable problem, however, casts some general doubt on the climate-economic approach. Due to its high degree

[15]Cf. Nordhaus, pp. 5 ff., but especially pp. 123 ff.

[16]Stern, *Review*; Welzer; Ekardt, *Theorie*, § 1; Kemfert, pp. 63 ff. (for an overview of pandemic events that are likely to happen see pp. 54 ff.).

[17]Parry et al. speak of 500 billion Euro total costs per year instead of only about 100 billion Euro.

[18]Cf. Stern, *Blueprint*, p. 39 and passim.

[19]Cf. Kemfert, p. 135 et seq.

[20]Thus, amiss Knopp and Piroch, pp. 409 ff. and Frenz, § 1 no. 1 ff. and passim; for a correct analysis see Wustlich, pp. 515 ff.

of complexity, climate change cannot exactly be predicted with respect to its concrete development and its economic impact. Rather, a high degree of uncertainty is immanent. Future uncertain events can hardly be integrated into precise cost calculations. For if a future event is not subject to specifiable probability (risk), but that probability is rather uncertain (uncertainty), this will render quantification impossible per se. Consequently, one cannot say something like "a looming damage of 10 billion Euro with a probability of 10% is valued at 1 billion Euro" in a case of uncertainty. Nor does this problem appear to be solved in the Stern Report, and as a result, Stern's critics draw the conclusion that rather low damage forecasts should be made.[21] However, a different conclusion (which is also a thesis of this study) could be more convincing: that the economic approach as a whole tends to suggest false precision and, therefore, what is necessary is a critical review as such.

Ultimately, these are all well-known problems – less with respect to climate data, but as regards economic loss amounts and dealing with uncertainty. Therefore, in the following, the focus will be more on other less-discussed problems of climate economics, which are not unique to Stern and the IPCC but also surface among their critics. The first concerns an almost entirely overlooked factual assumption – and then a set of normative assumptions, which are conditions for the further discussion, about whether the projected climate data and associated events (e.g. hurricanes or high oil prices) can lead to the classification as a "benefit" or a "loss".

3 The Limits to Growth

The most problematic factual assumption in climate-economic calculations of the "optimal climate policy" is the core assumption of "eternal" global economic growth – coupled with the focus on emission reductions to be achieved through technical measures (which is characteristic of the IPCC Working Group III). In this view of things, climate damage could perhaps result in (maybe significant) "setbacks in growth" (*Wachstumsdellen*). In the long run (!) however, after a recent economic revival due to the promotion of new technologies and after the (necessary) fight against poverty in parts of the world, an effective climate policy might require more of a critical revision of the idea of growth, something barely raised as an issue in the current climate-economic discussion. This also applies to Stern.[22] This problem is further reinforced by Stern and, apparently, the IPCC by accepting that climate change was a mere "market failure" (i.e. it is just seen as an *economic* and in the logic of current economics *solvable* problem).[23] Other economists such as Nordhaus fall far behind Stern and are even less open to critical questions on the validity of eternal growth.[24]

[21] On the following, in more detail, see Byatt et al., pp. 199 ff.

[22] E.g. Stern, *Blueprint*, p. 11 or p. 92; cf. also Weimann, p. 26.

[23] Cf. Stern, *Blueprint*, pp. 11 ff.

[24] Cf. Nordhaus, pp. 32 ff. and passim.

The cause of the climate problem is, however, in brief, the wealth of the industrialized world. When aiming at further growth, energy consumption and the consumption of fossil fuels also tend to increase. But climate protection at its core has to dramatically reduce the use of oil, gas and coal, and thus the amount of greenhouse gas emissions. Of course one can say: you can switch from fossil fuels to renewable energy – which emits only small amounts of greenhouse gases – and it is generally possible to use energy more efficiently.[25] These are key strategies to combat climate change. Thus, energy consumption, prosperity and economy seem to be able to continue growth, and yet shrink the greenhouse gas emissions. Climate protection is indeed a short-term opportunity for profits. For three reasons, however, sooner or later climate change will make it necessary to review the growth paradigm as such:

1. If economic growth continues without limit, the increase in wealth at least partly outweighs the greenhouse gas reductions from technically feasible energy efficiency and renewable energy on greenhouse gas savings ("rebound effect").[26] Figuratively speaking, if my car is still running ever more energy-efficiently, but globally more and more people drive a car (and I myself an ever bigger car), little is gained. And such is currently the trend. This explains why the emissions in developed countries have stagnated since 1990 despite various climate policy efforts.
2. If one wants to limit global climate change to a non-catastrophic level, *drastic* greenhouse gas reduction targets are urgently needed. It is not a matter of increasing global prosperity and keeping greenhouse gas emissions constant through greater efficiency or slightly decreasing them, but in fact it is necessary to reduce them globally (!) by about 80%. And these goals with the size of the challenge force us, besides "energy efficiency", also to contemplate an end to the paradigm of infinite growth in prosperity. For a reasonably stable global climate is the basis of human existence.
3. Ultimately something banal, but very basic applies: in a finite world, growth has physical limitations (unless we think of growth in education, piano-playing skills, etc.). It is impossible for the entire world – including all the Chinese, Indians, Indonesians, Brazilians, etc. who are gradually adopting the Occidental life-style and level of growth – to become infinitely richer. Even if humanity switches from fossil fuels to solar energy, other raw materials of this world remain finite. Wind turbines and eco-cars are made from resources, too. And that only "new ideas" grow permanently and thereby allow "eternal growth" without any use of resources, one might hope as much, but it seems an open question at best, so it is doubtful whether one should develop serious climate

---

[25]E.g. Stern, *Blueprint*, pp. 111 ff.

[26]The German Federal Environmental Agency found this effect to be true with regard to private energy consumption (cf. the underlying study available at: http://www.umweltdaten.de/publikationen/fpdf-1/3544.pdf.); even more pessimistic in this respect is the, albeit controversial, analysis by Garret.

policy recommendations on the basis of such an assumption.[27] As a general result, "ideas" potentially lead to the consumption of material resources. The Internet, for example, may seem an intangible idea, but computers and servers still require electrical power and finite and scarce resources for the production of various devices and the corresponding infrastructure.

All three problems are basic in kind. They cannot be negated by saying that the world today has, for example, larger oil reserves than was predicted 30 years ago. The problems can only be postponed (if that). The problem of "physical limitations" of the Earth also shows something essential: even without climate change, the common perspective on the idea of growth deserves a review.[28] This is also reflected at other points. Global growth rates, for example, give no information about the distribution of wealth: some are getting richer and those in need, who needed growth to occur the most, remain poor or get even poorer. Moreover, the growth concept – so far it is a well-known debate – ignores many aspects: private social work, such as private child care, and the ecological damage of the growth path to which there is currently deemed to be no alternative. Likewise, there is no empirical proof that growth per se increases human happiness.[29]

If however, the much-needed debate on climate change thus becomes a growth debate, this creates a serious problem. In most common economic views, capitalism and welfare need some form of growth, and even Marxists usually assume some form of growth. Whether this is so compelling is, of course, quite contentiously discussed.[30] The idea that a departure from the idea of growth would be the end of adequate human life seems at least historically dubious. For the whole of human history, up to the end of the eighteenth century, there were basically only static, i.e. non-growing economies.[31] Historically, a growth society is a special case tied to the occurrence of fossil fuels. Moreover, mankind in the age of fossil fuels has gained technical knowledge, which should nevertheless enable it to maintain the substantial achievements of this era.[32] Whatever one may think of this: the scale of the climate problem, the "rebound" and the physical finiteness of the world could

---

[27]The question of whether it would be reasonable or not to build modern societies on a vague prospective like that, is controversially discussed within the framework of the new Network on Sustainable Economy (www.nachhaltige-oekonomie.de). Even the Austrian government has spurred a discourse about the paradigm of unlimited growth (www.wachstumimwandel.at).

[28]Cf., on the following, the contributions by Schmidt; Behrens and Giljum; Löhr; Ekardt, 'Nachhaltigkeit', pp. 223 ff.; Daly; Wuppertal-Institut.

[29]Psychological research, however, implies the opposite; cf. Wuppertal-Institut, pp. 282 ff.

[30]Rogall, pp. 157 ff., tries to find an unbiased and balanced answer.

[31]Cf. Daly, passim. This alone hints at the fact that the idea of growth has a cultural background – which is not only rooted in classical liberalism alone, but also (already) in Calvinistic Protestantism; cf., with additional references, Ekardt, *Cool Down*, chapter II.

[32]The classic national "policy for growth and jobs" is further pressured by globalization and, accordingly, makes regulatory efforts even more difficult; see Ekardt, Meyer-Mews, Schmeichel and Steffenhagen, chapters 1 and 3.

spare any debate about it. To accept this, however, would mean that unlike the IPCC, Stern and much of the current research, one should no longer search solely for "new technologies", but (in the developed world) draw more on taking into account the possibility of sufficiency with regard to certain habits. Similarly, an increased reflection and research on the problems of a long-term "end of the idea of growth" would be appropriate.

One might ask whether a discussion on the previous problem be worthwhile. Who says that facts or forecasts of future facts on oil prices, hurricanes, etc., are of any relevance? Why do we not leave all this to the purely factual preferences of consumers? However, the present study is to oppose such a view. This leads to a review and critique of the preference approach, which is typical for economics – and also for the IPCC Working Group III with its mainly economic-engineering focus. At issue here are not just quantification and discounting, which are better treated separately (see below Section 5). It is rather a broader question of climate change and justice.

4 Climate Protection and Justice: Why Natural Science and Preference-Based Decision Theory Cannot Lay Sole Claim to Be Labelled "Objective" – and What "Ethics Within IPCC-Reports" Would Mean

4.1 The Core of Sustainable Ethics

This leads to a non-scientific (in the natural sciences sense), non-empirical but normative question, i.e. a question of "ought" or judgement: to what extent ought the (uncertain, but possibly drastic) negative and irreversible consequences, after a consideration of present interests, to be prevented or accepted? Because from an empirical stance, it does not necessarily follow that this observation is a welcome or unwelcome norm; even this basic fact is not sufficiently present in the economic and scientific debate.[33] This leads us into the field of ethics or theory of justice (the terms are used here interchangeably).[34] In the following, it will be shown that climate-economic models are not only based on questionable descriptive assumptions (see above) but also questionable normative and ethical assumptions. However, many economists would argue that their discipline has nothing to do with ethics at all when cost-benefit calculations or the efficiency of certain paths of possible climate policy are examined.[35] It will be shown below, however, that this is probably incorrect.

---

[33] Stern, *Blueprint*, pp. 86 ff. only hints at that problem and immediately forgets about it again.

[34] With regards to some of the possible misunderstandings that can arise in the context of the following Section see the controversy between Dilger, pp. 383 ff. and Ekardt and Kornack, 'Embryonenschutz', pp. 399 ff. (triggered by Ekardt, 'Verengungen').

[35] See, e.g., Wink; Nordhaus, pp. 175 ff.; Böhringer and Welsch, pp. 261 ff.

To show this, some more general thoughts are necessary. Let us assume the following thesis: a society is a just society if everyone can live according to their own ideas, and where all have a right to freedom, and conflicts of freedom are resolved through a democratic process, including a separation of powers. Human coexistence would be just if human rights or liberties, the rights to the basic preconditions of freedom, and certain other freedom supporting arrangements ("additional freedom conditions") respectively, were optimally achieved, including the ever-necessary balancing resolution of conflicts between the competing spheres of freedom. The considerations in the following sections will briefly try to show that this is the only necessary and possible criterion of justice, if one only interprets it right. Suffice it say that with a proper (re-) interpretation of democratic legal systems with respect to all of the following statements, there is convergence of a genuinely ethical and (in free democracies) a legal perspective, since human rights are the subject of international treaties and national constitutions.[36] The right to freedom is often referred to as human rights, which could be split up as general freedom of action, freedom of assembly, freedom of occupation, freedom to own property, religious freedom, freedom of speech, etc.[37] Legal and ethical traditions, however, often only parenthetically consider the protection of fundamental preconditions of freedom such as life, health, and subsistence (e.g. a basal access to energy, but also a sufficiently stable global climate[38]) as well as the freedom of future generations and people in other parts of the world. However, there is a strong argument that the protection of fundamental preconditions of freedom is already logically inherent in the concept of freedom itself – because without those basic preconditions there can never be freedom. An argument for the expansion of freedom in an intergenerational and global dimension will be given in Section 4.5. More detailed ethical and legal arguments for this "new" freedom – different from the classical liberal model of the West and in the spirit of freedom worldwide and for all generations – have been dealt with elsewhere.[39]

[36]Ethics is not only developing the principles of liberal democracies parallel to the law. In the following, it will be shown that it is also justifying them and, thereby, providing an ultimate basis for law; on the relation of law and ethics, see Ekardt, *Information*, § 1 A. (Law always combines normative and instrumental rationality).

[37]With regard to content, there is no further significance for this differentiation – apart from the idea that the constitutional lawmaker has in parts (pre-)structured the balancing of colliding freedoms (see Section 5) through deciding about their weight within the catalogue of fundamental rights.

[38]For the reasons to even include threatened damages (on a precautionary basis) that are not certain, see Section 2 above.

[39]For a detailed analysis of the theory of justice underlying Section 4 and with additional references, see Ekardt, *Theorie*, §§ 3–7; Ekardt, *Cool Down*, chapters 4–6; Ekardt 'Schutzpflichten'; focusing on the intergenerational dimension, see Unnerstall.

4.2 *A Key Distinction: Anthropology (Homo Economicus) Versus Normative Preference Theory/Efficiency Theory*

The important thing is: all these considerations are part of a theory of justice. In contrast, a theory of action describes the purely factual behaviour of humans, unlike a normative (moral or legal) consideration based on the theory of justice, which refers to how people ought to behave and how societies should be arranged. Instead of action theory one can also use the terms anthropology or the idea of man. Unfortunately much confusion is based on the common misunderstanding that an idea of man as erroneous was something normative, a picture of how man should be or how the society should be. This leads to a blurring of anthropology and theory of justice.[40] That the economists' common theory of action, which assumes that man is purely self-interested, is oversimplifying has been widely noticed in the last decades, though some economists might still use it. A theory of action stating that "man is purely factually (almost) only self-interested", i.e. Thomas Hobbes' theory of the homo economicus, is the focus of controversy in many discussions on economic theories. This doctrine, however, which helps economists to explain and forecast possible factual developments, will not be analysed here. Elsewhere it has been shown in this regard how companies, voters/consumers, and politicians are often linked in vicious circles to each other – and how factors such as conformity, emotional perception problems with spatiotemporal long-term consequences of own actions, self-interest, incorrect traditional values, technical-economic path dependency and structures of collective interests have thwarted drastic climate protection efforts.[41] Even though economic anthropologies do not always reach this necessary differentiation, their reference to the human tendency to self-interested behaviour makes a valuable contribution (in fact the concept of the homo economicus has correctly been modified in the last few decades and today is quite close to the ideas just presented).

Therefore, the real problem is not what Marxist economists often target: the empirically reasonably accurate descriptive anthropology of the *mainly* self-interested man. Neither is the problem a theory of happiness of life. With respect to the principle of freedom, such a theory of happiness lacks any general standards, so that there cannot be such a theory at all. Hence, an analysis of the dispute between some economists, who may see a particular increase in happiness as the result of economic struggle for profit, and their Marxist-inspired critics, who instead deem living a life of solidarity (as is allegedly a true human desire) happiness-increasing, is unnecessary on a theoretical level. In that regard, a freedom-based democratic

[40]This is very unfortunate as it causes a tendency to see facts in a somewhat askew, wishful way and as it, then, builds the basis for certain "do's and don'ts" – or, in contrast, as it prohibits to get through to the question whether solely normative statements can be justified at all. That is why, e.g., Heinig, pp. 330 ff. is missing the point.

[41]For references, see supra fn. 40. A slightly unsystematic list is offered in Rogall, pp. 63 ff. – who incorrectly assumes that sufficiency (i.e. "doing without" certain things and aspects of life) per se is voluntary, while instead it is much more often caused by (high) prices (see Section 6).

ethical and legal framework does not set any defaults, since there is no objective criterion for "happiness", and freedom allows no binding idea of happiness, either. However, a less "resource-focused" ideal of happiness would help many people recognize that their own freedom must be restricted for the sake of intergenerational and global freedom.

However, the problem is rather the (not only climate-) economics underlying the theory of justice, i.e. the efficiency theory or normative preference theory, as it is called in this context. Thus, the problem is not mainly descriptive anthropology but the normative theory of how human beings and societies *should* be.

4.3 Why the Economic Efficiency Theory (Normative Preference Theory) Is Ethics Itself – Also on the Concepts of Objectivity and Rationality

In order to show that an objective theory of justice is possible and that it must have the content that was briefly described above – and that the efficiency theory and normative preferences theory is a different but incorrect theory of justice – we first have to consider a question arising from the given arguments on freedom: is there a reason to deem the principle of freedom and its consequences (perhaps globally equal per capita emission rights) objectively just? Justice in this sense refers to the general rightness (*Richtigkeit*) of any social order. Therefore justice is not something "additional", which can be formulated after demanding prosperity or something similar. Any idea of how a society should be (even a simple "a society should be as rich as possible, and the distribution of wealth does not matter" or just "right is whatever the sum of the empirical preferences is"), is inherently a concept of justice, no matter whether it is right or wrong. Theories of a successful society – as can be found in works on moral philosophy, law, normative politics or moral theology – are per se concerned with justice, like physics or biology or sociology deal per se with a descriptive truth (even if some research might result in untrue findings, and therefore fails to meet the claim). The basic idea of neo-classical (including climate-) economists that it was necessary to maximize wealth, as expressed in valuable goods, is thus neither trivial nor even possible to classify as "empirical". This basic idea is, rather, a normative concept – it is an ethic (of efficiency),[42] which appears for the first time in Thomas Hobbes, like the homo economicus. Unlike anthropology it is not meant to explain or predict anything, but rather proposes the right decisions. It follows:

- "Efficiency versus justice" or "efficiency versus ethics" as an alternative, which economists like Stern or Nordhaus and their left-wing critics refer to, is just

[42] A striking analysis from an economic perspective is provided by Gawel, pp. 9 ff. and pp. 43 ff.

wrong.[43] The only reasonable discussion is whether the ethics of efficiency is right or wrong. Consequently, there is no point, if the IPCC in its fifth progress report wants to include an ethics or theory of justice analysis (the terms are synonymous) "in addition" to the efficiency analysis. This again erroneously assumes, that ethics (or justice) was a kind of diffuse part of the questions of social life, such as issues that seem somehow "very important" or even appear to have a religious connotations.[44]
- The controversy "ethics versus efficiency" rather concerns the question whether social equality in certain material goods as defined by increased redistribution should be reached to a greater extent. However, this is a more specific question. We come back to this briefly in Section 5.

But is there an objective ethics? Are there any objective, universal standards in a post-metaphysical, global, multi-cultural world (regardless of whether they are called "ethical" or "efficient")? That statements of fact, e.g. as regards anthropology or climate data, although partially uncertain and hard to prove, can be basically true and therefore objectively reasonable, i.e. rational, is seldom contested. Less clear is whether moral and legal norms may be correct and objective/rational. Many economists, including Stern, implicitly assume that only economic and empirical (natural) sciences can be rational. It will therefore be outlined briefly that there are indeed rational and objective norms and that freedom is their basic principle.[45] But first we have to define the following terminology:

- "Objective" means "not subjective", thus not subject to special perspectives, cultural backgrounds or settings, and therefore universal and valid everywhere.
- Reason, respectively rationality, refers to the ability to decide questions with reason, i.e. objectively. If it is a question of the rightness of moral and legal principles of justice – such as freedom and the rules for balancing conflicting freedoms that can be derived from it – then it is called normative reason (*normative Vernunft*). On the other hand, instrumental reason (*instrumentelle Vernunft*) and theoretical reason are related to facts. Instrumental reason is concerned with the question,

---

[43]That is why Nutzinger 'Effizienz', pp. 77 ff. and Grzeszick, pp. 647 ff. are, in my opinion, slightly misleading; see also Mathis, *Efficiency Instead of Justice?*.

[44]Unfortunately, the day-to-day and often even the scientific (if not philosophical) usage of the word "ethics" is rather arbitrary. It does not make any sense, however, to classify medically assisted suicide or the protection of embryonic stem cells as "ethical problems" and to leave out other societal questions that are normative as well (e.g. the scope of economic freedom).

[45]There are justification models that are (in parts) similar to the one we will develop here – yet, without the link to the questions of sustainability and climate protection. Cf. Alexy, *Recht*, pp. 127 ff.; Illies, pp. 129 ff.; Kuhlmann; Apel and Kettner; to some extent Habermas, pp. 56 ff.; implicit Ott and Döring, pp. 91 ff. The classics Immanuel Kant and John Rawls remain, in contrast, at least incomplete with regard to the theoretical basis of their substantiations although basic terms like rationality, human dignity, freedom, impartiality and separation of powers can be associated with them.

by what means can any norm which is assumed to be right e.g. a specific climate target (or a very selfish target such as theft) be implemented effectively, for example through emissions trading. Theoretical reason regards the determination of facts without any concrete related action, such as scientific climate research. Economist usually only accept the balancing aspect of normative rationality; the subject of this balancing, however, are preferences expressed in monetary values. That this is not convincing we will see in the further course of this analysis.

Whether there are objectively valid, i.e. rationally provable, norms and facts, is distinct from the – correct – observation that, factually, humans are often biased by subjective views when trying to determine facts and norms. This tendency for a subjective point of view is a natural one. But this by no means proves that objectivity – for example through careful examination and discussion with others – is altogether impossible.[46] We can consider the following example: it may be true that there are scientists who express their opinions for or against the presence of human-induced climate change because they expect financial benefits. Their statements were therefore not objective but subjectively distorted. But this does not mean that it is impossible to gain objective and unbiased insight on climate change. Furthermore, the finding that many perspectives are very "subjective" logically requires that there *are* objective perspectives – otherwise the subjective nature of those subjective perspectives could not reasonably be determined.

With respect to normative questions (unlike questions of fact) economists, sociologists, and political scientists mostly deny the possibility of objective statements altogether. For (not only climate-) economists, a "norm" is usually just what people prefer purely factually. "Rational" would apply only to quantifying (!) considerations, which transformed the not rationally verifiable preferences into a single "currency" (money) and thus made them comparable. If an economist asks for the right climate policy, he usually does not ask: what climate policy framework is set by freedom (including the freedom of those spatially and temporally far away, as well as balancing rules derived from freedom), under which various political decisions are possible? Economists would usually rather ask: how much would people living today be willing to pay for a stable global climate and what would be the advantages and disadvantages of climate change on the one hand and climate policy on the other hand, expressed in market prices? Such a preference theory might lead to the conclusion: what is valid is what everyone can agree on. Or: what is valid is the mathematical sum of preferences respectively expressed in money. Political scientists often tend to say: what is valid are simply the actual preferences of the respective majority.[47] It is important to note that in any case, even though these perspectives are founded on a theory of self-interested behaviour or anthropology

46 Berger and Luckmann, p. 2 have shown and advocated this differentiation in their classical (and often misperceived) analysis.

47 Often this is not expressed openly but presupposed implicitly; cf. only Stern, *Blueprint*, chapter 5; differing Ott and Döring, pp. 41 ff. and passim.

(homo economicus), as was mentioned above, they can still be considered as strictly separate. To put it bluntly, one can use the following simple formula: "People are in fact purely self-interested" (= anthropology) – "and this is a good thing, and listening to the purely factual preferences of the people is the best order of society" (= theory of justice, specifically the normative preference theory).

4.4 Why the Normative Preference Theory Is Not Convincing

The normative preference theory is the theoretical basis of how much climate change the respective economists deem objectively right and efficient respectively.[48] Any other approach, especially a normative argument without "figures", as will be developed in the course of this analysis, is usually declared unscientific and irrational. There are, however, strong objections against the preference theory not only but also with respect to climate protection:

- Quite familiar in neo-classical economics is the objection that the standard methods of identifying the actual preferences as numerical values simply do not work. The relevant issues and the necessary balancing of interests just cannot adequately be represented through prices. And it is impossible to detect actual preferences from real economic transactions on the basis of some kind of "disclosed morality of markets" (not even if de facto preferences as such were normatively relevant!). And even if this somehow were possible, most future damages cannot be discounted. This whole aspect of "simply not functioning" is the subject of a separate section (Section 5). Instead, it shall be shown here – and this might come as a surprise to economists – that regardless of those "application problems" the preference theory as such is unconvincing:
- According to the preference theory, our purely factual will is per se right (one could only ask whether the average utility, the sum of utilities or a genuine consensus shall be accounted for). Any normative test of "how the world actually is" is no longer present. The theory of justice or ethics as an independent discipline would be pointless and abolished per se.
- But we are not only facing a practical, but also a logical problem. For this is a naturalistic fallacy: why should our purely factual preferences ("is") be considered to be correct per se ("ought")?
- Furthermore, the question arises: should the factual ignorance as to the needs of future generations who cannot express any preference today per se be correct?
- If one pleads for majority rather than average preferences, there is the further question: whose preferences are meant? Can 50.1% of a society take any decision, or 73.4%, or 84.5%? And why should the majority per se always be right

[48] See, despite their contrariness, Stern, *Blueprint*, chapters 3 and 5, and Nordhaus, pp. 38 ff. and pp. 59 ff.

without any limits (as envisaged by a liberal democracy in the form of guarantees of freedom)?

- But particularly, the preference theory of justice entails a logical self-contradiction; for whoever says that there are no general normative propositions, and therefore general preferences should be determinative, makes a general statement about norms. The statement "everything is relative with respect to norms" refutes itself. The possibility of objective morality just cannot logically be contested. Its denial contradicts itself.
- It should be noted that all these arguments also apply with respect to some kind of ethics that is not based explicitly on preferences, but goes something like: "what is just, is the society that represents the current de facto national traditions."

All this does not mean, of course, that self-interested preferences, for example – or de facto national traditions – do not play a major role for the *factual enforcement*, i.e. the governance of climate protection. It was only pointed out that a *normative (moral or legal) justification* of climate protection – or a normative limitation or refutation – cannot be based on those preferences. But the principle of freedom, including its rules of balancing might be suited for this purpose. This principle can take future generations into account, is not subject to any of the types of problem just described, while retaining the basic intention – everyone should be able to determine themselves – and deriving it compellingly.

4.5 The Case for a Theory of Justice Based on Discourse Rationality as a Better Alternative to the Preference Theory

However, this is correct only under a major condition: namely, if the principle of freedom, including all principles derived from it, is the basis for the universal standard for justice. But why should this be right? And why should such a statement possibly be "objective"? We can briefly consider the following: in a pluralistic world we necessarily argue on normative issues. Even fundamentalists and autocrats do so inevitably, at least *occasionally*. And they avail themselves of the human language. But who argues with reasons (i.e. rational, that is with words like "because, since and therefore"), who uses phrases such as "X is valid because of Y" and with respect to normative questions logically assumes (1) the possibility of objectivity in morality, and (2) the existence of freedom – whether he wants it de facto or not:[49]

[49] So called negative or transcendental pragmatic arguments of the following kind have been used by Alexy, *Recht*, pp. 127 ff.; Illies, pp. 129 ff.; Kuhlmann, passim; implicitly also Ott and Döring, pp. 91 ff. and passim. The structure of a negative (and not deductive) argument with which an infinite regress or a "randomly chosen axiom" can be prevented goes back to Plato, Augustine and Thomas Aquinas (as a logical figure but not referring to the issue at hand). For some misunderstandings that often occur in the "philosophy/economy" discourse, see the dispute between Dilger, pp. 383 ff. and Ekardt 'Umweltpolitik', pp. 399 ff.

1. We imply logically that normative questions can be decided using reasoning in general and ergo objectively and not only subjectively, preference based, otherwise we contradict ourselves. We assume this (a) even every day when we pose normative theses and justify them, that is to say we attach them with the claim of objective acceptability (rather than to present them only as subjective). And it would be almost impossible never to use words such as "because, since, and therefore" with respect to normative questions. Thus, there is no escape from the fundamental *possibility* (!) of objectivity in normative issues. We even logically imply the possibility of objective statements (b) if we say: "I am a sceptic, and say there are objectively only subjective statements about morality." This statement can only be valid if there is objectivity. Thus, the criticism raised towards objectivity voids itself.
2. We also logically imply that potential discourse partners deserve equal impartial respect – because reasons are egalitarian and the opposite of violence and degradation, and they are addressed to individuals with intellectual autonomy, because without autonomy one cannot assess reasons. No one could say: "My theory X and its reasons could easily be refuted by Mr P, but you, Mr Q, as a fool, should believe in it." And no one could say: "After we had P silenced we finally were able to convince ourselves that Y is a good reason for X." It therefore contradicts the very meaning of "reasons", to understand the act of reasoning as relative to the person being addressed – a reason is convincing and can be tested by anyone. Someone who gives reasons in a conversation about justice (i.e. uses sentences with "because, since and therefore"), but then disputes the other's respectability ergo contradicts what he assumes logically.

This means: logically, whoever engages in the dispute of justice based on reason must respect the partner as an equal – regardless of whether he is aware of the implications of his reasoning or whether he intends to reason only to persuade the other, for it is all about the strictly *logical implications* of our speech (but not about our purely *factual self-image* which per se does not imply anything). The respect for autonomy as self-determination as required by reason must apply to the individual and therefore mean respect for individual autonomy: collectives as such are in fact not possible discourse partners. This is rather the individual human being arguing.[50]

This is the justification for the principle of respect for the autonomy of individuals (human dignity[51]). In addition, but hardly distinguishable, this also founds

[50] A whole set of fictive or real arguments against this justification of (1) the possibility of rationality and (2) of human dignity and impartiality as sole universal principles that can be deducted from rationality are discussed by Ekardt, *Theorie*, § 3; Ekardt, *Demokratie*, chapter 3.

[51] The principle of human dignity itself is not a freedom right or human right. It is not a norm at all that refers to any kind of singular case, either ethical or legal. Human dignity is rather the reason for human rights (in contrast to being a norm/a right on its own); it, therefore, guides the application of other norms – in our case, different types (realms) of freedom that belong to human beings – and proclaims autonomy as the central idea of our legal system. The "inviolability" of human dignity and its visible – in norms like Art. 1 par. 2–3 of the German constitution and the EU Charter of

the principle that justice means independence from subjective perspectives (impartiality). From in turn this follows the right to freedom for all people[52] – and only the principle of freedom: due to the lack of compelling reasons, other principles cannot interfere with the principle of freedom. Therefore, the same freedom based self-determination, along with its supporting preconditions, is the sole criterion of justice. Being man in general, after all, requires necessarily (only) the right to self-determination for all. And this right to freedom applies to all people, even if I never talk to them. For reasons in issues of justice (unlike statements made in private or aesthetic issues) are addressed to anyone who could potentially disprove them – therefore, I have to recognize all people are to be respected, as soon as I *occasionally* use reasons, and everyone does. This is made clear by the following control example. No one could seriously say: "The absent Mr P could immediately refute my theses – but because of your stupidity you should believe them." This, of course, is not valid reasoning.

The principle of freedom is thus universally founded. And because all potential discourse partners are included, as we have just seen, I must also concede freedom to people living spatially and timely far away. This is (a) one of the key arguments for the extension of the principle of freedom to future generations and therefore for global justice and intergenerational justice and hence for sustainability – and also (b) to the idea that freedom as such, implies protection exactly at the point at which freedom is threatened. A "Kantian discourse ethics" concept of reason and autonomy, as outlined here, in this case would opt differently from a "economic-Hobbesian" concept. However, both concepts are concerned with freedom, but for the discourse of ethics, not just in the sense of consumer sovereignty and factual consumer preferences.[53]

5 The Balancing Processes – Efficiency Through Quantifications and Discounting?

Solving the generational and global conflict between many competing freedoms, i.e. determining the right amount of climate policy, is not an easy task. Both, the normative weighing or balancing of interests and the relevant facts (see Section 3

Fundamental Rights – character as a "reason" shows that all this is not only philosophically but also legally correct. For the current state of discussion, see Ekardt and Kornack 'Embryonenschutz', pp. 349 ff.; Ekardt and Kornack 'Gentechnik'; similarly, e.g., Enders; for a contrasting viewpoint, see Böckenförde, 'Menschenwürde', pp. 809 ff.; differentiating, Heinig, pp. 330 ff. and pp. 353 ff.

[52]That freedom exists because of dignity is, e.g., explicitly stated in Art. 1 par. 2 of the German constitution which says "darum" (= therefore) exists freedom, i.e. because of human dignity, and is also supported by the explanatory documents (*Gesetzgebungsmaterialien*) on the EU Charter of Fundamental Rights; see Ekardt and Kornack, 'Gentechnik'.

[53]Although following a different path, this is also the conclusion of Rothlin and Ott and Döring, pp. 78 ff. and pp. 91 ff.; rather a (in my opinion hardly to the point) critique of profit-oriented competition can be found in Hoffmann, pp. 23 ff.; see further, Nutzinger, 'Effizienz', pp. 7 ff. and pp. 51 ff.

above), which are necessary to ascertain in how far a certain normative concern is actually affected, are characterized by uncertainty. The problems encountered with regards to the climate facts have already been discussed above. It is also possible to ethically and similarly legally derive rules for the balancing of interests from the principle of freedom and infer institutions of balancing (as has been done elsewhere in more detail[54]). An example of a rule of balancing would be that the factual basis of a decision has to be determined as carefully as possible.[55] Another rule is that only freedom and the (broadly understood) freedom conditions are possible concerns that are relevant for balancing. Another one is that freedom and its fundamental and "further" conditions may only be interfered with as far as it is necessary to strengthen other freedoms and freedom conditions. Another rule – again, already inherent to the very concept of freedom itself – promulgates that if someone is to be obliged *ex ante* to prevent or *ex post* to remedy an impairment of a freedom, this should, wherever possible, be the causer of the impairment. Earlier in this study we derived another rule, namely the precautionary principle: even under uncertain circumstances, the interference with freedom or its conditions need be recognized, but possibly with less importance attached. Many other rules can be derived. In all this there generally is no "one correct" result when balancing interests. This is true for climate policy as well. Consequently, there is certain leeway with respect to a just climate policy – but this is not arbitrarily large. And the bodies which have to use this leeway within the framework of the balancing rules are also not arbitrary: rather, an institutional rule can be derived from freedom saying that a decision maker, who can be elected and deselected has to make the decision. Where necessary, further specifications must be made by authorities and courts observing the principle of the separation of powers; furthermore, there must be constitutional courts to verify compliance with the balancing rules.[56]

Economists, however, quantify all interests concerned and calculate what the "right" level of climate protection is. Everything that has a value for people, i.e. where a respective factual preference exists, should be translated into monetary terms, including life and health – or it should be disregarded.[57] Specific rules of balancing are unnecessary within the framework of such an approach. The facts of benefits and harm merge with the preferences. This sounds attractive insofar as no leeway is required – theoretically "exactly one" policy recommendation can be made and the results are "exact figures". This, however, is problematic in several ways. Firstly, (see Section 4 above) the underlying normative preference theory in itself is not convincing. Secondly, (see Section 3) benefits and damages, which

[54]Cf. supra fn. 40; similarly Susnjar and Alexy, *Grundrechte*.

[55]The actual decision of a certain extent of climate protection policy based on the weighing and balancing of interests or efficiency thinking, is itself a normative statement and not a factual one. Facts alone can never deliver decisions as they are only possible if normative criteria are available.

[56]Furthermore, one can deduce that there should be a decision on the national or transnational level, whichever is suited best for it (the global level in case of climate protection policy); see Ekardt, Meyer-Mews, Schmeichel and Steffenhagen, chapters 1, 3 and 5.

[57]Cf. Nordhaus, p. 4; critical also Burtraw and Sterner.

already have a market price, lack sufficiently precise facts if, as with climate change, the entire world economy is involved with unmanageable numbers of individual actions, and also periods of more than 100 years. Thirdly, there are, as already indicated and now further demonstrated, more insurmountable problems of application of the preference theory:[58] the calculation of climate change costs (and, in comparison, climate policy costs) disguises the fact that essential concerns cannot be quantified in monetary terms, e.g. (massive) damage to life and health.[59] The absence of damages to life and health due to climate change has no market price, neither has peace in the sense of "absence of conflicts over resources." Thus, both cannot reasonably be quantitatively used to offset the economic effects of climate change and climate policy. Neither can an artificial market price be determined for concerns without an actual market price, as economists keep doing through the "hypothetical willingness to pay" for life and health, i.e. the absence of hurricanes, wars, etc. This is already true since this willingness is fictitious and therefore not very informative (that assigning preferences based on a "morality of markets" does not help is discussed in the analysis of the discounting method below). Moreover, the willingness to pay is of course limited by the ability to pay and would lead to the remarkable result that, for example, Bill Gates' interests are worth much more than a Bangladeshi's, because Bill Gates can pay a lot more and the Bangladeshi can pay nothing. This was also noted by Stern, contrary to the economic mainstream, and yet he too suddenly uses monetary values for "non-market effects".[60] If he allocates the same amount for every human, that would in fact be correct (see below), but in the context of the preference theory it is without justification and therefore inconsistent.

Another problem of climate economics is discounting:[61] future damages are said to weigh less than today's. This is understandable, at least superficially, if the victim today and in ten years is the same person. But why should a Bangladeshi's damage in 50 years (1) per se be less important than my damage today? One could say: future people cannot express any preferences today, so they are uninteresting. This idea is, as has been indicated, inherent in the preference theory. But then, consequently, one would not have to discount, but to completely disregard the damages of someone who is not yet alive. And compared to those living today the discounting is inconsistent with regard to the passage of time. Given the preference theory, why should an economist be allowed to dictate whether I have a present preference and should not care for the future? The expectation of perpetual growth (2) also cannot justify discounting, whether with respect to those already living today or to future generations. Let us not forget the limits of growth. Also, (3) the empirical observation

[58]Cf. Ekardt, *Theorie*, § 5 C.; Mathis, pp. 113 ff.; Otsuka, pp. 109 ff.; Meyer, pp. 136 ff.

[59]For a critique concerning this matter see Ekardt, *Theorie*, § 5; in parts also Mathis, pp. 113 ff.; Otsuka, pp. 109 ff.; Meyer, pp. 136 ff. On this issue, see in the present volume Balz Hammer, 'Valuing the Invaluable?'.

[60]Conceding to this is Stern, *Blueprint*, p. 92.

[61]For a detailed and critical analysis of the problem of discounting, see Unnerstall, pp. 320 ff.; cf. also Rawls; supporting the method of discounting is Birnbacher. On this issue, see in this volume Klaus Mathis, 'Discounting the Future?'.

of real market prices ("morality of markets"), which according to many economists expresses the preference for the present over the future, does not justify discounting. For (a) there are no observable market or interest-rate developments that would say anything about what factual preferences exist in terms of damages over several centuries – and with irreversible character. Moreover, (b) drawing conclusions from market prices, only considers the preferences of today's people.

These preference determinations based on a "morality of markets" are criticized by Stern (stating this as a criticism against most other economists),[62] but not the growth-oriented discounting. Stern certainly offers an argument for discounting which is at least worth considering: (4) the uncertain probability of future losses. However, whether this can be mathematically expressed is doubtful. At least where no mathematical probability can be determined, a supposedly clear discount rate is ultimately arbitrary, and therefore is not superior to general balancing rules as were introduced above. And even if all this could be disregarded, discounting would only be possible if the respective damage could actually be expressed in monetary terms despite the above criticism. And this is often not the case.

All this shows once again the fundamental problem of (not only, but especially climate-) economic approaches: behind seemingly clear mathematical results, assumptions are concealed which are far from universally compelling, but are rather contestable in important respects. This criticism is not limited to normative assumptions (e.g. to discounting and the preference theory) but is also directed at factual assumptions: e.g. on the extent of looming climate damage or the growth idea. *Hence, it is impossible to calculate the correct amount of climate protection and the associated distributional issues required by morality and legal principles.* Rather, it is necessary to make climate-policy decisions within the limits set by the described rules of balancing – worldwide and nationally. As repeatedly indicated, such a decision must mean more climate protection than previously. Briefly stated:[63] (1) the existing climate policy probably already disregards the balancing rule that its decisions must be based on correct facts: in particular, the recent actions are probably erroneously deemed suitable to avoid the looming of drastic damage caused by climate change. (2) Furthermore, politics so far has not taken into account, in its decision-making, that the basic right of freedom also has an intergenerational and a global cross-border dimension and therefore the legal positions of future generations and the proverbial Bangladeshis have to be considered in parliamentary/legal decisions.[64] (3) The human right to a subsistence minimum as an elementary precondition for freedom (which is not only a right of those living here and now, but is also an intergenerational and global right) can be overcome in balancing only

[62]Cf. Stern, *Blueprint*, pp. 80 ff. and pp. 95 ff.

[63]On a legal and ethical level that also implies: in case of actions against lawmakers constitutional courts have (or had) to decide in favour of the plaintiff and force lawmakers to rethink and re-decide on their respective climate protection policy with the following aspects in mind. In more detail, cf. Ekardt 'Schutzpflichten'.

[64]Focusing less on the preventive level and (in my opinion sub-optimally) more on the subsequent level of liability is Verheyen.

in limited areas, because freedom is pointless without this physical basis. But this right must also include access to basal energy and at least a somewhat stable global climate, which in turn requires drastic climate policies. This, too, has currently not been taken into account by decision-makers. Similarly, no consideration has been made for the limited remaining emissions budget equal distribution in view of (a) its scarcity, and (b) the imperative nature of low emissions for human survival.[65] An egalitarian distribution is also proposed by Stern, but with the mistaken reason (relying on the uncertainty of the burden of proof) that there were ultimately no grounds to argue against an equal distribution.[66] It should be mentioned once again that all this is meant as both an ethical and a legal statement.

To verify the factual basis of a political decision, economic research is undoubtedly extremely valuable – and it is also helpful for balancing to the extent that goods with a market price are concerned and unpolished figures are generated which also account for, e.g., the costs of possible climate wars (this is not included in the Stern Report[67]). If a calculation is done, one should at least try to include all the real monetary costs to the extent that they are recognizable. In this way, economists can provide crucial factual material for balancing – within the framework of the overall balancing theory. It shows for instance that the actual monetary damage to the climate, such as crop failures or other weather damages, would be more expensive than an effective climate policy. These are key benefits of the IPCC reports and the Stern Report. Equally important are statements on the probabilities of events. In my view, however, economists and natural scientists can often only provide those probabilities with a lower degree of accuracy than one would expect. The natural conditions of climate change and the global economy are simply too complex. A perhaps more modest model that is not normative, but also less quantifiable and less focused on natural science – a climate economics which is merged with the other climate social sciences within the framework of a balancing theory – could be a feasible alternative. Provided, however, that climate social science is concerned with these themes: limits of growth, a normatively and logically rigorous theory of justice, a theory of

---

[65]With regards to ideas on a substantial climate change policy, including a (virtual) per-capita-distribution of emission-rights as the basic criterion for "climate justice" (with some modifications concerning the problem of the industrialized countries' historical emissions), see Ekardt, *Cool Down*, Chapter 4 and 5; Ekardt and von Hövel, pp. 102 ff.; this is economically presupposed – and without any real normative justification – by Wicke, Spiegel and Wicke-Thüs, and (albeit without citing them and a number of other authors) WBGU 2009.

[66]The approach developed here, in contrast to Sen, has justified (and not only asserted) universal freedom and, therewith, the relevance of its preconditions (and, furthermore, a theory from which rules structuring the balancing process can be deduced). These advantages also exist compared to "theories concerning basic human needs" (inspired by Marxist or Rousseauian ideas); in addition, the latter also have the flaw of mixing descriptive anthropology and normative theory of justice. Also, they do not have a method to determine its basic categories (what is there a "need" for?) and they mingle justice and conceptions of what a "good life" is supposed to look like (with potentially authoritarian tendencies). Viewed against this background, Ott and Döring, pp. 78 ff. seems to be problematic.

[67]Stern, *Review*, p. 151, only very generally speaks about increasing "instability".

balancing, anthropology, also a governance and control theory which is based on more than purely economic perspectives (see Section 6 below).[68] Also in governance, climate economics is and remains very important, but again not exclusive. It is therefore a welcome development that Stern admits the omissions of the economic approach – if only in general and without addressing the basic problems of growth and preference theory.[69]

On the other hand, the efficiency theory must be defended against John Rawls' accusation stated under the (once again) misleading heading "efficiency versus justice." Rawls criticizes that the efficiency theory – in other words, the utilitarian and Hobbesian ethics – does not recognize *absolute* rights, i.e. rights that cannot be offset by other rights, not to be confused with *universal* rights meaning "applicable everywhere"!).[70] Even though this is true for the efficiency theory, just as it is for the balancing approach advocated in this study, given the many possible collisions of freedoms which are at the heart of (climate) policy, there is little need to do so. Absolute guarantees of freedom are only rarely justifiable, mainly when balancing would undermine the liberal character of the system as a whole (for example torture in order to convict criminals).

Until now, some key points on climate change and justice conflicting with the dominant climate economics can be summed up as follows: ethical (including climate-ethical) findings are not empirical and, especially, not natural-scientific observations; they are rather normative (= judgement/ought) findings. Even though the application of an ethical or legal norm often refers to scientific (factual) questions, these facts do not infer as such any ethical or legal result. Furthermore, the basic principles of ethics, although normative in their nature, can objectively be specified. Ethics is not "subjective" or "mere convention", and is not founded on "axioms" with arbitrary starting points. On the other hand, the actual decision of specific ethical issues is somewhat blurry. Yet, the balancing rules and the institutional competences limiting the discretion are again objective. Since ethics is generally concerned with the conflict between different interests, every ethical decision is ultimately a balancing problem between different conflicting freedoms (and their preconditions). Absolute obligations or strict balancing prohibitions (e.g. an absolute right to environmental stability at any price which cannot be balanced with other interests) are ethically and legally hard to justify.[71] This does not mean that

---

[68]Many climate social scientists, however, favour working on merely factual descriptions of existing (and possibly incorrect) theories of justice, climate discourses, how climate is perceived and so on – cf., in this respect, some of the articles in Voss – which seems to be less important (unless it is helpful to elucidate the anthropology behind inadequate climate protection).

[69]Cf. Stern, *Review*, pp. 149 ff.

[70]Cf. Rawls, p. 19. German legal scholars – e.g. Böckenförde, *Staat*, pp. 188 ff. – tend to make the same mistake and seem to think that rejecting quantifications would also include the dismissal of balancing procedures (in most cases). Thereby, they mistake the universality of values for their absoluteness. See also Heinig, pp. 353 ff., who does not distinguish precisely between the principles referring to justice and the subsequent balancing procedure.

[71]Based on what was demonstrated here, one could also try to give an answer to the question whether the often-repeated accusations that economic efficiency analysis turns a blind eye to

the balance can be resolved by a mathematical quantification – even though "figures" have the advantage with respect to politics and the media that they allow complex statements to be easily displayed. Therefore "figures", even if they represent a new welfare index, as defined by Amartya Sen and others, for the "landmark gross national product" (the latter being calculated on the basis of valuable goods) as is currently being discussed in France, can only be symbols, but no replacement for complex balancing.

6 Governance: Can "More Business Ethics and CSR" Be Effective Climate Protection Instruments? On the Misleading Separation of "Bottom Up" and "Top Down" Approaches

To conclude this paper, one last question shall be raised briefly: what conclusions do economists draw from efficiency analysis or from the balancing process with regards to climate policy instruments? Elsewhere I have supported and further developed the idea of a worldwide emissions trading system, which is also pursued by many economists, however, based on much more drastic climate protection goals and with a dual social component within the industrialized countries and with respect to the developing countries as a compensation for global and strict climate protection goals.[72]

The fact that the proposed approach has to work on a global level, follows (a) from the global nature of the climate problem and (b) from the threat of a simple shift of emissions from a country with ambitious climate policy into another country (carbon leakage) which would be devastating for both climate protection and competitiveness – for example, if steel companies transferred their industrial plants from Europe, to China. However, there are economists who seem to focus on "bottom up" approaches on climate protection instead of political regulations, i.e. on voluntary corporate climate protection activities. Certainly any voluntary corporate commitment in terms of climate protection (or sustainability in general) is welcome. For the company itself, this should be attractive, either as a means of customer acquisition, or to motivate employees, or simply as a means of cost savings (e.g. with respect to resource consumption). However, appeals to individual companies or citizens,

questions regarding distributive justice are correct. This answer would probably be: yes and no. Because, there is no way to deduce a strict imperative that says we have to redistribute extensively. Certain "social elements" result from theory of justice with respect to the balancing procedure, like a right to a subsistence minimum; beyond that the lawmaker has a wide margin for questions of distributive justice. Cf. Ekardt, Heitmann and Hennig, and Ekardt, *Demokratie*, chapter V.

[72] For further details, see the references supra fn. 40.

and a reliance on their voluntary initiatives, unregulated free trade, and industry self-regulation[73] cannot replace binding climate policy regulations.[74]

- Firstly, the individual citizen or entrepreneur is not the appropriate authority always to undertake the ethically necessary complex balancing of different interests. This should rather be the task of politics formed into a legal order, i.e. the legislature. The problem of "too little specificity" is a standard problem of purely ethical appeals, if they are not transposed into a legal form and thus substantiated.
- There is a second fundamental problem when relying on purely voluntary activity: this will often only work as far as potential property interests of the company are involved. And when a massive change is needed, the precise question is: can we really expect that, for example, the auto industry will "voluntarily" (i.e. without economically incentivizing instruments such as emissions trading) adapt the social model "only car-sharing" and will therefore switch to the production of bicycles? Why should the mostly self-interested man, who is regularly diagnosed by economists, reduce emissions to almost zero on a purely (!) voluntary basis? And how will rebound effects from companies' private pursuit of growth disappear, if they might try to produce more efficient products but ultimately want to sell more products than before? And how can consumers, especially in light of economists' demand for realistic anthropology, be truly expected to exert pressure for the described necessary change through their purchasing decision? Especially as those worst affected by climate change, the world's poor and the future poor, have the lowest purchasing power to exert market pressure on companies through their purchasing decisions. Ultimately an entrepreneurial initiative always remains a variant of the general growth paradigm – which is doubtful.

In that regard, on an instrumental or governance level we must adhere to the anthropological insights of many "climate macro economists" as opposed to CSR-oriented climate micro economists: climate appears on the market superficially as a "free" good and is therefore used too heavily. And there are many other human characteristics such as short-term interest, the tendency to convenience and habit, the emotional non-perception of spatial-temporally remote loss, etc., which further exacerbate the problem. The only response is the creation of regulations (such as taxes or certificate markets), which provides clear enforcement mechanisms and sanctions for the given targets and which already today put a price on impending climate damage and thus stop the "market failure". That this, so far, is only occasionally compared to the challenges can be explained with the described "vicious circle" of politics and voters. However, this does not change the fact that without political and legal regulations, which in turn, due to the vicious circle, depend on a

[73] As an example for the following problems, see Becker, pp. 7 ff.; Davidson, pp. 22 ff.; Wieland; Suchanek and Lin-Hi, pp. 67 ff.

[74] In more detail and with further references, see Ekardt, *Information*, § 1 C. II.; Ekardt, *Theorie*, § 7.

social rethink, a solution to the climate problem cannot be expected. All this cannot be changed by demanding a general "bottom up" rather than "top down" approach to climate policy. Of course, voluntary actions ("bottom up") are welcome in principle. But where they cannot be expected with reasonable certainty, other alternatives are required. One cannot argue that this approach is adverse to freedom. Precise political regulation considerably protects the freedom of future generations as well as the people in transition and developing countries which have contributed little to climate change.

Instead, economic preference theory is destabilizing modern democracy: the seemingly exact climate-economic statements make politicians appear completely irrational if they do not follow the climate policy proposed by economists. They are not. Therefore, the other climate social sciences should no longer leave the leading role to climate economists. Not only in the interest of climate protection but also in the interest of a further improved climate economics that, at first glance, might appear more humble but ultimately integrates a more convincing and realistic concept for weighing and balancing interests.

Bibliography

Alexy, Robert, *Theorie der Grundrechte* (Frankfurt a.M., 1986; cited as: *Grundrechte*)

Alexy, Robert, *Recht, Vernunft, Diskurs* (Frankfurt a.M., 1995; cited as: *Recht*)

Apel, Karl-Otto and Kettner, Matthias (ed.), *Zur Anwendung der Diskursethik in Politik, Recht und Wissenschaft* (Frankfurt a.M., 1993)

Baumert, Kevin A., Herzog, Timothy and Pershing, Jonathan, 'Navigating the Numbers, Greenhouse Gas Data and International Climate Policy', from *World Resource Institute* (2005)

Becker, Gerhold, 'Moral Leadership in Business', in *Journal of International Business Ethics*, Vol. 2, No. 1 (2009), pp. 7 ff.

Behrens, Arno and Giljium, Stefan, 'Der globale Ressourcenabbau', in *Forum für angewandtes systemisches Stoffstrommanagement*, Vol. 3 (2005), pp. 13 ff.

Berger, Peter and Luckmann, Thomas, *The Social Construction of Reality. A Treatise in the Sociology of Knowledge* (Garden City NY, 1966)

Birnbacher, Dieter, *Verantwortung für zukünftige Generationen* (Stuttgart, 1988)

Böckenförde, Ernst-Wolfgang, *Staat, Verfassung, Demokratie* (Frankfurt a.M., 1991; cited as: *Staat*)

Böckenförde, Ernst-Wolfgang, 'Menschenwürde als normatives Prinzip', in *Juristenzeitung*, Vol. 58 (2003), pp. 809 ff. (cited as: 'Menschenwürde')

Böhringer, Christoph and Welsch, Heinz, 'Effektivität, Fairness und Effizienz in der internationalen Klimapolitik: Contraction and Covergence mit handelbaren Emissionsrechten', in Franz Beckenbach et al. (eds.), *Diskurs Klimapolitik, Jahrbuch Ökologische Ökonomik* Marburg, (2009), pp. 261 ff.

Burtraw, Dallas and Sterner, Thomas, *Climate Change Abatement: Not "Stern" Enough?* (2009), available at: http://www.rff.org/Publications/WPC/Pages/09_04_06_Climate_Change_Abatement.aspx

Byatt, Ian et al., 'The Stern Review: A Dual Critique. Part II. Economic Aspects', in *World Economics*, Vol. 7 (2006), pp. 199 ff.

Daly, Herman, *Beyond Growth. The Economics of Sustainable Development* (Boston, 1996)

Davidson, Kirk, 'Ethical Concerns at the Bottom of the Pyramid. Where CSR meets BOP', in *Journal of International Business Ethics*, Vol. 2, No. 1 (2009), pp. 22 ff.

Dilger, Alexander, 'Ökonomik versus Diskursethik. 10 Thesen zu Felix Ekardt', in *Zeitschrift für Umweltpolitik und Umweltrecht*, Vol. 29 (2006), pp. 383 ff.

Ekardt, Felix, 'Verengungen der Nachhaltigkeits- und Umweltschutzdebatte auf die instrumentelle Vernunft – am Beispiel der Wirtschaftswissenschaften', in *Zeitschrift für Umweltpolitik und Umweltrecht*, Vol. 27 (2004), pp. 531 ff. (cited as: 'Verengungen')

Ekardt, Felix, 'Ökonomik versus Diskursethik in der Umweltpolitik: Antikritische Bemerkungen zu Alexander Dilger', in *Zeitschrift für Umweltpolitik und Umweltrecht*, Vol. 29 (2006), pp. 399 ff. (cited as: 'Umweltpolitik')

Ekardt, Felix, *Wird die Demokratie ungerecht? Politik in Zeiten der Globalisierung* (Munich, 2007; cited as: *Demokratie*)

Ekardt, Felix, *Cool Down. 50 Irrtümer über unsere Klima-Zukunft – Klimaschutz neu denken* (Freiburg, 2009; cited as: *Cool Down*)

Ekardt, Felix, 'Nachhaltigkeit und Recht', in *Zeitschrift für Umweltpolitik und Umweltrecht* (2009), pp. 223 ff. (cited as: 'Nachhaltigkeit')

Ekardt, Felix, 'Schutzpflichten, Abwägungsregeln, Mindeststandards und Drittschutz', in *Die Verwaltung*, Beiheft 11 (2010), pp. 27 ff. (cited as: 'Schutzpflichten')

Ekardt, Felix, *Information, Partizipation, Rechtsschutz. Prozeduralisierung von Gerechtigkeit und Steuerung in der Europäischen Union* (2nd. edn., Münster, 2010; cited as: *Information*)

Ekardt, Felix, *Theorie der Nachhaltigkeit. Rechtliche, ethische und politische Zugänge* (Baden-Baden, 2010; cited as: *Theorie*)

Ekardt, Felix, Exner, Anne-Katrin and Albrecht, Sibylle, 'Climate Change, Justice, and Clean Development. A Critical Review of the Copenhagen Negotiation Draft', in *Carbon & Climate Law Review*, Vol. 3 (2009), pp. 261 ff.

Ekardt, Felix, Heitmann, Christian and Hennig, Bettina, *Soziale Gerechtigkeit in der Klimapolitik* (Düsseldorf, 2010)

Ekardt, Felix and Kornack, Daniel, 'Embryonenschutz auf verfassungsrechtlichen Abwegen?', in *Kritische Vierteljahreszeitschrift für Gesetzgebung und Rechtswissenschaft*, Vol. 89 (2006), pp. 349 ff. (cited as: 'Embryonenschutz')

Ekardt, Felix and Kornack, Daniel, '"Europäische" und "deutsche" Menschenwürde und die Gentechnik-Forschungsförderung', in *Zeitschrift für europarechtliche Studien*, Vol. 13 (2010), pp. 111 ff. (cited as: 'Gentechnik')

Ekardt, Felix, Meyer-Mews, Swantje, Schmeichel, Andrea and Steffenhagen, Larissa, 'Globalisierung und soziale Ungleichheit – Welthandelsrecht und Sozialstaatlichkeit', in *Böckler-Arbeitspapier* Nr. 170 (Düsseldorf, 2009)

Ekardt, Felix and von Hövel, Antonia, 'Distributive Justice, Competitiveness and Transnational Climate Protection: "One Human – One Emission Right"', in *Carbon & Climate Law Review*, Vol. 3 (2009), pp. 102 ff.

Enders, Christoph, *Die Menschenwürde in der Verfassungsordnung* (Tübingen, 1997)

Frenz, Walter and Müggenborg, Hans-Jürgen (ed.), *Kommentar zum Erneuerbare-Energien-Gesetz* (Berlin, 2009)

Garrett, Tim *Are There Basic Physical Constraints on Future Anthropogenic Emissions of Carbon Dioxide?* (2009), available at: http://www.springerlink.com/content/9476j57g1t07vhn2/

Gawel, Erik, 'Ökonomische Effizienzanforderungen und ihre juristische Rezeption', in id. (ed.), *Effizienz im Umweltrecht* (Baden-Baden, 2001), pp. 9 ff.

Grzeszick, Bernd, 'Lässt sich eine Verfassung kalkulieren?', in *Juristenzeitung*, Vol. 58 (2003), pp. 647 ff.

Habermas, Jürgen, *Moralbewusstsein und kommunikatives Handeln* (Frankfurt a.M., 1983)

Hänggi, Marcel, *Wir Schwätzer im Treibhaus. Warum die Klimapolitik versagt* (Zurich, 2008)

Hansen, James E., 'Scientific Reticence and Sea Level Rise', in *Environmental Research Letters*, Vol. 2, No. 2 (2007), 6 pp.

Heinig, Hans Michael, *Der Sozialstaat im Dienst der Freiheit. Zur Formel vom "sozialen" Staat in Art. 20 Abs. 1 GG* (Tübingen, 2008)

Hoffmann, Johannes, 'Ethische Kritik des Wettbewerbsrechts', in Johannes Hoffmann and Gerhard Scherhorn (eds.), *Eine Politik für Nachhaltigkeit. Neuordnung der Kapital- und Gütermärkte* (Erkelenz, 2009), pp. 23 ff.

Illies, Christian, *The Grounds of Ethical Judgement – New Transcendental Arguments in Moral Philosophy* (Oxford, 2003)

IPCC, *Climate Change 2007. Mitigation of Climate Change* (2007)

Kemfert, Claudia, *Die andere Klima-Zukunft* (Hamburg, 2008)

Knopp, Lothar and Piroch, Ingmar, 'Umweltschutz und Wirtschaftskrise – Verschärfung des Spannungsverhältnisses Ökonomie/Ökologie?', in *Zeitschrift für Umweltrecht*, Vol. 20 (2009), pp. 409 ff.

Kuhlmann, Wolfgang, *Reflexive Letztbegründung* (Freiburg and Munich, 1985)

Löhr, Dirk, 'Zins und Wirtschaftswachstum', in *Forum für angewandtes systemisches Stoffstrommanagement*, Vol. 3 (2005), pp. 33 ff.

Lomborg, Björn, *Cool it! Warum wir trotz Klimawandel kühlen Kopf bewahren sollten* (Munich, 2007)

Mathis, Klaus, *Efficiency Instead of Justice? Searching for the Philosophical Foundations of the Economic Analysis of Law* (New York, 2009)

Meyer, Kirsten, 'How to be Consistent without Saving the Greater Number', in *Philosophy & Public Affairs*, Vol. 34 (2006), pp. 136 ff.

Nordhaus, William, *A Question of Balance. Weighing the Options on Global Warming Policies* (New Haven, 2008)

Nutzinger, Hans, 'Effizienz, Gerechtigkeit und Nachhaltigkeit', in id. (ed.), *Regulierung, Wettbewerb und Marktwirtschaft, Festschrift für Carl Christian von Weizsäcker* (Göttingen, 2003), pp. 77 ff. (cited as: 'Effizienz')

Nutzinger, Hans G. (ed.), *Gerechtigkeit in der Wirtschaft – Quadratur des Kreises?* (Marburg, 2006; cited as: *Gerechtigkeit*)

Otsuka, Michael, 'Saving Lives, Moral Theory, and the Claims of Individuals', in *Philosophy & Public Affairs*, Vol. 34 (2006), pp. 109 ff.

Ott, Konrad and Döring, Ralf, *Theorie und Praxis starker Nachhaltigkeit* (Marburg, 2004)

Parry, Martin et al., *Assessing the Costs of Adaptation to Climate Change: A Review of the UNFCCC and Other Recent Estimates* (2009), available at: http://www.iied.org/climatechange/key-issues/economics-and-equity-adaptation/costs-adapting-climate change-significantly-under-estimated

Posner, Richard, 'Wealth Maximization Revisited', in *Notre Dame Journal of Law, Ethics and Public Policy*, Vol. 2 (1986), pp. 85 ff.

Rawls, John, *A Theory of Justice* (Cambridge MA, 1971)

Rogall, Holger, *Nachhaltige Ökonomie* (Marburg, 2009)

Rothlin, Stephan, *Gerechtigkeit in Freiheit – Darstellung und kritische Würdigung des Begriffs der Gerechtigkeit im Denken von Friedrich August von Hayek* (Frankfurt a.M., 1992)

Schmidt, Matthias, 'Wachstum mit Zukunft', in *Forum für angewandtes systemisches Stoffstrommanagement*, Vol. 3 (2005), pp. 7 ff.

Sen, Amartya, *Development as Freedom* (Oxford, 1999)

Stehr, Nico and von Storch, Hans, Anpassung und Vermeidung oder von der Illusion der Differenz., in *GAIA*, Vol. 17 (2008), pp. 19 ff.

Stern, Nicholas, *Stern Review Final Report*, available at: http://www.hm-treasury.gov.uk/stern_review_report.htm (2006; cited as: *Review*)

Stern, Nicholas, *A Blueprint for a Safer Planet: How to Manage Climate Change and Create a New Era of Progress and Prosperity* London, (2009; cited as: *Blueprint*)

Suchanek, Andreas and Lin-Hi, Nick, 'Unternehmerische Verantwortung', in Rupert Baumgartner, Hubert Biedermann and Daniela Ebner (eds.), *Unternehmenspraxis und Nachhaltigkeit* (Munich and Mering, 2007), pp. 67 ff.

Susnjar, Davor, *Proportionality, Fundamental Rights, and Balance of Powers* (Leiden, 2010)

Unnerstall, Herwig, *Rechte zukünftiger Generationen* (Würzburg, 1999)

Verheyen, Roda, *Climate Change Damage and International Law: Prevention Duties and State Responsibility* (Leiden, 2006)

Voss, Martin (ed.), *Der Klimawandel. Sozialwissenschaftliche Perspektiven* (Wiesbaden, 2010)

Weimann, Joachim, *Die Klimapolitik-Katastrophe* (Marburg, 2009)
Welzer, Harald, *Klimakriege* (Frankfurt a.M., 2008)
Wicke, Lutz, Spiegel, Peter and Wicke-Thüs, Inga, *Kyoto Plus* (Munich, 2006)
Wieland, Josef, *CSR als Netzwerkgovernance* (Marburg, 2009)
Wink, Rüdiger, *Generationengerechtigkeit im Zeitalter der Gentechnik* (Baden-Baden, 2002)
Wissenschaftlicher Beirat Globale Umweltveränderung and WBGU, *Kassensturz für den Weltklimavertrag. Der Budgetansatz* (Berlin, 2009)
Wuppertal-Institut, *Zukunftsfähiges Deutschland in einer globalisierten Welt* (Frankfurt a.M., 2008)
Wustlich, Guido, 'Ökonomisierung im Umweltrecht', in *Zeitschrift für Umweltrecht*, Vol. 20 (2009), pp. 515 ff.

Valuing the Invaluable?

Valuation of Human Life in Cost-Efficiency Assessments of Regulatory Interventions

Balz Hammer

Starting from the practical necessity of valuing human life for the purposes of assessing the cost-efficiency of state projects aimed at the mitigation of risk (healthcare, environmental protection, etc.), this essay begins by outlining and offering a critical appraisal of the various economic methods. In addition to this theoretical presentation of the methodology, it is exemplified by discussing the findings and practical significance of several important studies on the value of a human life, finally leading on to the controversial debate on the permissibility of placing an economic value on human life. The focus of interest is the question of whether and how any monetarization of human life can be justified from an ethical and legal viewpoint within the constellations described. The conclusion draws together the insights gained.

1 Introduction

An important aspect of sustainable development, particularly in the field of environmental protection, is the mitigation of risk such as preparedness for natural disasters. As a consequence of general scarcity of resources, state authorities see themselves forced to examine public projects prior to realization to determine their cost-efficiency. The instrument used in practice to assess efficiency is cost-benefit analysis.[1] If human lives are affected by the project in question – e.g. by interventions in the field of environmental protection, or indeed in preventive healthcare or transport safety – the decision-maker has no choice but to assign a value to the human lives saved or sacrificed by this project: this means weighing the costs of the project to be assessed against the value of the human lives affected. The outcome of this cost-benefit analysis depends on the per-capita amount at which each saved or

[1] Spengler, p. 270.

B. Hammer (✉)
University of Lucerne, Faculty of Law, Frohburgstrasse 3, P.O. Box 4466, CH-6002 Lucerne, Switzerland
e-mail: balz.hammer@unilu.ch

K. Mathis (ed.), *Efficiency, Sustainability, and Justice to Future Generations*, Law and Philosophy Library 98, DOI 10.1007/978-94-007-1869-2_10,

sacrificed life is valued. For life-saving interventions, placing a higher value on life increases social utility while a lower value decreases utility.[2]

Prompted by this, economists have conducted numerous studies on the valuation of human life.[3] These studies are referred to by agencies such as the US Environmental Protection Agency (EPA) when assessing the cost-efficiency of planned regulations. In relation to the Clear Skies Act of 2002, the EPA put the value of each human life to be saved by this legislation at US$ 6.1 million.[4] This compares with the view taken by the Swiss Federal Office for the Environment (BAFU, formerly BUWAL) in respect of gravitational natural hazards that a threshold value of CHF 10 million is "appropriate" for rescue costs, where rescue costs are understood to mean the costs of precautions to save one (statistical) human life.[5] By clearly and openly declaring the cost-benefit considerations used in the valuation of the human lives saved by the intervention in question, these regulatory authorities create transparency, which ultimately contributes to the legitimacy of the planned project.

In order to illustrate how these values are arrived at, the economic analysis methods used for the valuation of human life will be assessed and critically appraised in the following section.

2 Economic Methods for the Valuation of Human Life

Measurement of the value of a human life can essentially be undertaken from two different viewpoints: either through the eyes of the affected person himself or herself, or through the eyes of other people (e.g. family members, or the whole of society).[6] Taking these two perspectives as starting points, two economic methods for determining the value of human life can then be distinguished: the *willingness-to-pay method* (subjectively objectivized) and the *human-capital method* (objective). Whereas the human-capital method mainly comes into play in liability law, nowadays, where it is used for quantifying damages for the loss of a breadwinner, in the area of regulatory interventions it has been superseded by the

[2]Viscusi, 'Devaluation', p. 103.

[3]For example the life of a citizen of the United States has been valued at US$ 7 million on average. Viscusi, 'How', p. 311.

[4]United States Environmental Protection Agency, *Technical Addendum: Methodologies for the Benefit Analysis of the Clear Skies Initiative* (2002), pp. 35 f., downloadable from: http://www.airimpacts.org/documents/local/Tech_adden.pdf (viewed on: 18 October 2010). This figure attracted particular public attention because it was only deemed to apply to people aged under 65, while a lower value was assigned to people over 65 years old. This age-based differentiation will be examined in more detail later, see Section 3.2.1 below.

[5]Bundesamt für Umwelt, Wald und Landschaft (BUWAL), *Risikoanalyse bei gravitativen Naturgefahren, Methode,* Umwelt-Materialien Nr. 107/I (Bern, 1999), pp. 107 f., downloadable from: http://www.bafu.admin.ch/publikationen/publikation/00131/index.html?lang=de (viewed on: 18 October 2010).

[6]Merten, p. 29.

willingness-to-pay method.[7] In the following, these two methodological approaches will be discussed and critically appraised.

2.1 Human-Capital Method

Under this method, the value of a human life is equivalent to the value of the human capital destroyed by the killing of a human being.[8] In this analysis, human capital is understood to mean the totality of all human skills, capabilities and knowledge which have a positive influence on the income of the person concerned.[9] Therefore activities such as vocational training or a university education are considered to be investments in an individual's human capital.

2.1.1 Method for Quantifying the Value of Destroyed Human Capital

A human life cannot be valued with reference to the cost of the aforementioned investments since these are an indicator of the *acquired* but not the total *destroyed human capital.* To determine the latter, the definitive measure is the – future, hypothetical – lifetime income and tax dues for the individual in question. Making reference to these two criteria, two different approaches for measuring the destroyed human capital can essentially be distinguished.[10]

The first possibility is the discounting of a person's future lifetime income to the time of his or her death. The resulting value is generally known as the present discounted value or cash value. If one proceeds to subtract from this the expected future consumer expenditure of the deceased person, one obtains what is known as the *discounted net value of lost income.*[11] This is the main figure used by the US civil courts to calculate the value of damages for the loss of a breadwinner in cases where someone has been killed.[12]

Another approach to determine the destroyed human capital is based on the sum of the tax payments that would have been due from the deceased person in the future. If all payments from the state to the deceased (e.g. transfer payments, subsidies, etc.) are duly taken into account, the figure is called the *foregone net financial contribution to society.*[13] One obvious shortcoming of this approach is that – unlike

---

[7]Mishan and Quah, N 1 to § 36; Viscusi, 'How', pp. 311 f.

[8]Merten, p. 30.

[9]Mathis, 'Generationen', p. 196; Viscusi, 'Value', p. 589.

[10]For the sake of simplicity and clarity, only the two most important approaches to the human-capital method are presented here. For a more detailed account of the different variants of the human-capital method see Mishan: 'Evaluation of Life and Limb: A Theoretical Approach'.

[11]Landefeld and Seskin, p. 556.

[12]Viscusi, 'Survey', p. 214; Viscusi, Harrington and Vernon, p. 721.

[13]See Schelling, pp. 138 ff.

the first – it does not take account of payments from the victim's income to himself or to his dependents.[14]

In summary, it is clear that the value of a (concrete) human life under the human-capital method corresponds to the human capital destroyed by the killing of the individual concerned, i.e. depending on the underlying approach, either the discounted net value of the income lost as a result of the individual's death or the foregone net financial contribution to society. From this, the underlying idea behind this valuation method is clear: an individual is always of value as long as he is productive throughout his lifetime.[15]

2.1.2 Critical Appraisal of the Human-Capital Method

An important advantage of the human-capital method is primarily in its straightforward application, since the data needed in order to calculate it are relative easy to obtain.[16] Nevertheless, both approaches to the human-capital method have various shortcomings, which will be discussed briefly in the following.

The main problem, initially, is that this method offers no solution for valuing the lives of those people who do not yet generate added value for society (e.g. children) or no longer do so (e.g. the retired or people with an occupational incapacity).[17] Although it can be assumed that those *not yet* economically active are likely at some point to pursue paid employment, earn an income and pay tax contributions to the state, it remains unclear on what hypothetical basis the relevant calculations should be made.[18] In contrast, those *no longer* employed or now incapacitated have earned an income and paid taxes before ceasing to work. Under the human-capital method, however, only their future payments are relevant. Since these cease at the point of retirement or incapacity, members of this group no longer generate any added value for society – with the exception of any taxation on capital. As a consequence, under the human-capital method, the value of their life would equal zero;[19] clearly an absurd result.

Furthermore, the human-capital method largely disregards those services that are performed unremunerated (e.g. housework, voluntary services or honorary roles), since strictly speaking these neither generate an income nor make any financial contribution to society.[20] From an economic point of view, however, such services very much contribute to maximizing social welfare and thus to "enhancing the value" of the individual in question, which is why from a welfare-economic perspective they should also be given due consideration.

---

[14] Viscusi, Harrington and Vernon, p. 721.

[15] Landefeld and Seskin, p. 556.

[16] Landefeld and Seskin, p. 557.

[17] Landefeld and Seskin, p. 556; Merten, p. 31.

[18] Nida-Rümelin, p. 903.

[19] Ackerman and Heinzerling, p. 72; Landefeld and Seskin, p. 556.

[20] Landefeld and Seskin, p. 556; Merten, p. 31.

Another problem in calculating destroyed human capital is the discounting rate, i.e. the interest rate at which future payment streams are discounted to the time of death in order to calculate the present discounted value.[21] As yet this discounting rate is highly contested, and is therefore applied very inconsistently by the relevant authorities.[22] This inconsistency not only causes confusion but can also result in abuse if the authority concerned can elect to use whichever interest rate will produce the result it desires.[23]

Furthermore, the value of life varies greatly depending on the victim's occupation, since the sole basis for the calculation of destroyed human capital is the individual's income or tax dues. Accordingly, a banker is of greater "value" than a fire-fighter. Although this differentiation appears absurd, in liability law it is still the standard approach for calculating the damages for the loss of a breadwinner who has been killed.[24] For this is ultimately a matter of compensation for loss between two parties, in other words a compensatory payment by the defendant for the harm suffered by the claimants as a result of the loss of their breadwinner. For this reason, differentiations of this kind based on the individual's income or tax dues should continue to be permissible.[25]

The final point to be made is that the human-capital method is far more concerned with the *costs of death* (foregone income and medical expenditure) than the value of a human life.[26] The question asked is not what value life has or had for the person killed, but what costs the victim's relatives or society as a whole suffered due to his or her death. Therefore the human-capital method is nowadays practically only used in liability law to quantify the damages for loss of a breadwinner. In contrast, in the area of state regulatory interventions – as a basis for assessing the cost-efficiency of a concrete project – it has been superseded by the willingness-to-pay method for measuring the value of a human life.[27]

2.2 Willingness-to-Pay Method

As mentioned earlier, due to the general scarcity of resources, even the state cannot avoid examining the cost-efficiency of public projects in advance of their realization by means of a cost-benefit analysis. In this process, economists work on the

---

[21] Landefeld and Seskin, p. 556.

[22] Viscusi, 'Survey', p. 214. Essentially, however, one should discount at the current market interest rate. On this issue, see in the present volume Klaus Mathis, 'Discounting the Future?'.

[23] Landefeld and Seskin, p. 557.

[24] Ackerman and Heinzerling, pp. 64 f. and 71; Viscusi, 'Survey', p. 214.

[25] Nevertheless from an economic analysis of law perspective, the question that arises is whether this practice of the courts and the resultant sometimes low damages payments do not set the wrong incentives with regard to the avoidance of harmful events. On this point, see the case study in Viscusi, 'Survey', pp. 215 f.

[26] Viscusi, 'How', p. 313; id., 'Survey', p. 208.

[27] Viscusi, 'Survey', pp. 197 f. and 214.

assumption that the value of the utility gain from a public regulatory intervention can be ascertained by establishing how much those who benefit from the resulting utility would be willing to pay. Where the project in question is intended to reduce mortality risks, i.e. the utility of the intervention in question is the avoidance of a certain number of deaths, the value of the lives saved corresponds to the willingness to pay of those individuals who benefit from the risk reduction.[28] For this reason, in the applied area of the cost-benefit analysis of public regulations, the willingness-to-pay method is now the method of choice for calculating the value of a human life. The elements of this procedure will be discussed in detail below.

2.2.1 Starting Point and Aim of the Willingness-to-Pay Method

The starting point of the willingness-to-pay method is the fact that from day to day, people in different areas of life are exposed – consciously or unconsciously – to certain risks. Guido Calabresi comments on this acceptance of risk in today's highly technologized society in the following terms:

> Our society is not committed to preserving life at any cost. In its broadest sense, the rather unpleasant notion that we are willing to destroy lives should be obvious. [...] [L]ives are spent not only when the *quid pro quo* is some great moral principle, but also when it is a matter of convenience. Ventures are undertaken that, statistically at least, are certain to cost lives. [. . .] We take planes and cars rather than safer, slower means of travel. And perhaps most telling, we use relatively safe equipment rather than the safest imaginable because – and it is not a bad reason – the safest costs too much.[29]

Due to the general scarcity of resources, all individuals find themselves forced to weigh up costs and benefits to decide whether or not certain precautionary measures to reduce everyday risks are worthwhile. Accordingly, people are subject to a trade-off – in economic parlance – between money and risk, because they have to decide whether they either wish to take a certain risk or else bear the costs of reducing it.[30]

The willingness-to-pay method ties in with these risk analyses observed on the individual decision-making level by inquiring into the value implicitly attached to an individual's own life.[31] This is not a simple matter of asking individuals what their lives are worth to them – since the answer would probably be "An infinite amount!".[32] Instead, the willingness-to-pay method deploys a variety of scientific analyses to determine the value of human life, either with reference to the individual's willingness to pay for measures to reduce the risk of death (known as the

---

[28]Viscusi, 'Survey', p. 197.

[29]Calabresi, pp. 17 f. In a similar way, Guido Calabresi had described the risk society in an earlier essay, 'The Decision for Accidents: An Approach to Nonfault Allocation of Costs'.

[30]Ashenfelter, p. C10.

[31]When individuals buy a car and spend more money on additional safety features, or when they sacrifice the chance to earn more pay for accepting higher risks of a workplace accident, they are expressing their willingness to pay for the reduction of mortality risks, which equates to an implicit valuation of their life.

[32]Griffin, p. 53; Nida-Rümelin, pp. 888 f.

willingness-to-pay approach) or with reference to the compensation demanded by the individual for acceptance of the risk of death (known as the *willingness-to-accept* approach).[33] These studies focus not on concrete deaths but on future actual mortality risks. What is crucial is that these are always specific but at the same time small probabilities.[34] In contrast to the human-capital method, the aim is not to calculate the value of a particular person's life but the life of an indeterminate average person. Hence, mention is often made in this context of the *value of a statistical human life*. This "value of a statistical life" (VSL) is ultimately derived by dividing the willingness to pay by the mortality risk, i.e. the level of risk reduction purchased,

$$\textit{Value of a statistical life} = \frac{\textit{Willingness to pay}}{\textit{Size of risk reduction}};$$

or by dividing the compensation claim by the heightened level of risk accepted, i.e. the risk premium paid as a component of salary,

$$\textit{Value of a statistical life} = \frac{\textit{Willingness to accept}}{\textit{Size of risk increase}}.$$

This formula clearly reiterates that the VSL is based on a "money-risk trade-off", i.e. a balance struck between money and small risks of death.[35]

2.2.2 Methodology for Determining Willingness to Pay

In order to arrive at the willingness to pay or the compensation claim figure, in practice recourse is taken to a variety of scientific analysis methods, which can essentially be differentiated according to two different approaches to data-gathering: under the *"stated-preferences methods"* subjects are questioned directly as to their willingness to pay, whereas under the *"revealed-preferences methods"* willingness to pay is derived from the market behaviour of economic subjects.[36] In the following, the most important analytical procedures developed with reference to these two approaches will briefly be presented, in each case citing a fictional example in order to illustrate the VSL calculation.[37]

2.2.2.1 Contingent-Valuation Method

Under this stated-preferences method, a large number of subjects are questioned directly about how much they would pay for a reduction of the mortality risk from

---

[33] Merten, p. 179.

[34] Viscusi, 'Value', p. 586. On the groundings of this principle, see Schelling, pp. 142 ff.

[35] Viscusi, 'How', p. 312.

[36] A brief and clear presentation of the various methods of ascertaining willingness to pay and its practical significance is found in Brannon, pp. 60 ff.

[37] These constructed examples lay no claim to factual accuracy and serve solely to illustrate the method for valuing human life.

1/10,000 to 1/100,000. To calculate the value of a statistical life, the mean sum of money stated in these subjects' responses is divided by the size of the risk reduction; i.e. for an average willingness to pay of EUR 500, the value of a human life amounts to EUR 500: (1/10,000) = EUR 5 million.[38]

The advantage of direct questioning is, firstly, its great simplicity and flexibility as a technique, which makes it applicable to a variety of regulatory domains. Secondly, the targeted nature of the inquiry makes it possible to ascertain the individual willingness to pay to reduce the risks associated with specific activities, which are almost impossible to determine effectively by means of revealed-preferences methods since the necessary market data are often unavailable.[39] However, one basic problem attaches to this method, which makes it unattractive to many practitioners: questions are only ever *hypothetical in nature*, meaning that the subjects' responses may not reflect their behaviour in a real situation.[40] Furthermore, the manner of questioning or the questioning technique can substantially influence the assessment of risks (a bias known as the *"framing effect"*). Overall, only limited reliance can be placed on information obtained by this method about actual willingness to pay and for use in calculation of the VSL.[41]

2.2.2.2 Consumer-Market-Behaviour Method

This analysis method focuses on the balancing of risks by consumers between the mortality risk associated with a product, and its price (a "money-risk trade-off"). Thus, it ascertains consumers' willingness to pay for the safety of the products they purchase, i.e. the amount that consumers are willing to pay in order to reduce the risk of death associated with the purchased product (*willingness-to-pay* approach).[42] The following example briefly illustrates this method.

On the assumption that out of 10,000 car drivers, one will die as a consequence of a traffic accident, each individual driver is statistically confronted with a mortality risk of 1/10,000. Let us further assume that by installing an airbag at a cost of EUR 500, each of the drivers in question can reduce his or her mortality risk from 1/10,000 to 1/100,000. All those who are prepared to pay this amount put a value on their own life of at least EUR 500: (1/10,000) = EUR 5 million.

2.2.2.3 Labour-Market-Behaviour Method

The labour-market-behaviour method is the method most frequently applied in practice for determining the VSL.[43] Here, economists compare the salaries of jobs

---

[38]See Brannon, p. 61.

[39]Ashenfelter, p. C11; Viscusi, 'Value', p. 586.

[40]Ackerman and Heinzerling, p. 81.

[41]Brannon, p. 61; Viscusi, Harrington and Vernon, p. 718.

[42]Viscusi, 'Value', p. 589.

[43]Viscusi, 'Survey', p. 202.

which are, as far as possible, identical but which carry different levels of mortality risk.[44] The factor of central interest is the "compensating wage differential", i.e. the salary premium[45] that must be paid in order to persuade the employee – *ceteris paribus* – to accept the risk of death associated with the more dangerous work. Under this analysis, the salary premium can be understood as compensation for the increased probability of a fatal occupational accident.[46] The labour-market-behaviour method is prompted by the idea that the employee, in accepting this salary premium, simultaneously expresses that he is reconciled to the increased risk of death (*willingness-to-accept* approach), and thus implicitly places a certain value on his life.[47] Precisely how this value is determined is illustrated by the following example.

Imagine that there is a choice between two comparable jobs, X and Y. The annual salary for job X is EUR 75,000, whereas Y is remunerated at EUR 80,000 per year. In the case of job Y, however, the annual risk of a fatal occupational accident is 0.1% higher than for job X. Now, anyone who chooses job Y implicitly places a value on his or her life of EUR 5000: (1/1000) = EUR 5 million. As this example makes clear, the employee concerned has to choose between a higher salary or a lower risk of death. In the context of the labour-market-behaviour method, this is referred to as a *"wage-risk trade-off"*.[48]

According to a meta-analysis conducted by Viscusi and Aldy in 2003,[49] the VSL for half of the labour-market-behaviour analyses conducted in the USA lies somewhere in the region of US$ 5–12 million, while the median for all the studies reviewed is US$ 7 million (stated in year-2000 prices).[50] In Switzerland, Barranzini and Ferro Luzzi also carried out a labour market study on the calculation of the VSL in the year 2001, using two different labour market datasets in conjunction with sector-specific data on fatal occupational accidents. According to their calculations

---

[44]Brannon, p. 60.

[45]Examples of this kind of salary uplift for a risk-laden job (known as a risk premium) are allowances paid to police officers or firefighters. In practice, however, such risk premiums are not always made explicit, since age, training and experience also play an essential role in the level of salary. Therefore further research steps are often necessary in such studies in order to identify the precise salary-risk trade-off. See Spengler, p. 273.

[46]Ackerman and Heinzerling, pp. 75 f.; Spengler, p. 271. The compensatory salary differentials approach can be traced back to the following statement by Adam Smith: "The wages of labour vary with the ease or hardship, the cleanliness or dirtiness, the honourableness or dishonourableness of the employment." Smith, p. 100. Thus, Smith assumed that a worker had to be compensated for accepting an additional risk in the workplace.

[47]On the exact procedure of labour market analysis, particularly on determining the compensatory salary differentials, see Viscusi, 'Survey', pp. 203 ff.

[48]Viscusi, 'Value', p. 587.

[49]By means of a meta-analysis, the results of various (primary) studies on the VSL are summarized and analysed, in the attempt to control or equalize exogenous factors that have a possible influence on the level of the calculated VSL (such as the differences in risk of the different studies). See Brannon, pp. 61 f.

[50]Viscusi and Aldy, p. 18.

and depending on the underlying labour market dataset, the VSL in Switzerland lies between US$ 6.08 million (Swiss Labour Force Survey of 1995) and US$ 8.31 million (Swiss Wages Structure Survey of 1994), both at 1999 prices.[51] Converted into prices for the year 2000, this equates to US$ 6.3 million and US$ 8.6 million.[52] According to the study, therefore, the VSL in Switzerland is approximately in line with the results of the US studies.

2.2.3 Critical Appraisal of the Willingness-to-Pay Method

Like the human-capital method before it, the willingness-to-pay method is also subject to numerous conceptual and practical problems which have not yet been resolved in a satisfactory way. The most important of these problems will be discussed briefly below.

2.2.3.1 Problem of Bounded Rationality

The concept of "bounded rationality"[53] expresses the fact that in situations where there is a clear choice, human beings behave rationally; in situations involving more difficult choices, however, the rationality of our behaviour is very limited.[54] As we have seen, the willingness-to-pay method always places the central focus on hypothetical and very small mortality risks. The individuals concerned often have major difficulties in correctly quantifying the scale and size of these probabilities[55] so that psychological distortions can arise in the perception and assessment of risks: due to the inadequate quantity and quality of the available information, risks which are actually small are often overestimated while risks which are actually large are underestimated.[56] This faulty assessment of the actual mortality risk ultimately leads to an *impaired expression of the willingness to pay or willingness to accept*, which in turn distorts the calculation parameters used to determine the value of a statistical life.[57]

Other mistaken perceptions in the assessment of risk can also arise because of the nature of the mortality risk in question. For example, there is found to be a significantly higher willingness to pay to reduce the risk of a terminal cancer illness

[51] Baranzini and Ferro Luzzi, p. 158.

[52] Viscusi and Aldy, p. 28, table 4.

[53] On the groundings, see Herbert A. Simon: 'A Behavioral Model of Rational Choice'.

[54] Merten, p. 179.

[55] Ashenfelter, p. C16; Broome, p. 263.

[56] Sunstein, 'Plea', p. 427; Merten, pp. 180 f. On the theme of psychological distortions, particularly in the judicial adjudication process, see in the present volume Klaus Mathis and Fabian Diriwächter, 'Is the Rationality of Judicial Judgements Jeopardized by Cognitive Biases and Empathy?'.

[57] Ackerman and Heinzerling, p. 77.

than to reduce the risk of a sudden and unexpected accidental death.[58] Moreover, willingness to pay is also influenced by the time of realization of the mortality risk. For often, the greater the future remoteness of the eventual occurrence of death, the lower the willingness to pay for the reduction of that risk.[59]

2.2.3.2 Problem of Risk-Aversion

Another problem of the willingness-to-pay method stems from the fact that, in most cases, the relevant analyses only cover those people who actually take the risk of mortality, whereas individuals who do not do a dangerous job or who refuse a job offer with a high mortality risk tend to be excluded from data collections.[60] This is predominantly because jobs and occupations with a greater mortality risk are often preferred by people who are less risk-averse (a phenomenon referred to as "cool-headedness"[61]), who for their part require comparatively lower levels of compensation for taking the risk in question.[62] Aside from these, however, another group of people exist who are fundamentally risk-averse and are only prepared to enter into the same level of mortality risk at substantially higher prices, and perhaps not at any price. This heterogeneity in acceptance of risk among individuals is often given too little attention, particularly in labour-market-behaviour analyses, which often record only those people who have a greater risk-affinity or, at least, lower risk-aversion than others.[63] As a consequence of this overrepresentation of individuals with lower risk-aversion, the average willingness to pay or willingness to accept *calculated* on the basis of the analyses is generally lower than the *actual* value. Therefore the calculated VSL needs to be corrected upward accordingly.[64]

2.2.3.3 The Agency Problem

Other weaknesses of this method come to light when determining the willingness to pay of jobs or occupations in which the individual concerned is not alone in

[58]On this phenomenon of the additional willingness to pay to reduce the risks of dying in pain and suffering in the event of a so-called "bad death" see Sunstein, 'Deaths', pp. 268 f.

[59]Merten, pp. 209 f.

[60]Ashenfelter, p. C12; Viscusi, 'Value', p. 588.

[61]Accordingly, in dangerous occupations, employees are systematically selected for special capacities to deal with risky situations – e.g. particular level-headedness or adroitness. Garen, p. 10.

[62]Spengler, p. 271 and p. 274; Sunstein, 'Plea', p. 412. Women are much more risk-averse than men, which is why they comparatively rarely do high-risk jobs or activities. On this, see Ackerman and Heinzerling, pp. 78 f.

[63]Ashenfelter, p. C16.

[64]Ackerman and Heinzerling, pp. 77 f. It can be objected, however, that lower risk-aversion in certain cases need not necessarily result in a lower VSL. On this ambivalent relationship between risk-aversion and the VSL see Eeckhoudt and Hammitt: 'Does risk aversion increase the value of mortality risk?'.

benefiting from a risk reduction, since third parties also stand to gain.[65] On the one hand, risks of this kind with third-party effects can result in opportunistic behaviour on the part of certain individuals, who remain passive because they assume that others also subject to the risk will pay to reduce the risk. These so-called free riders would be willing to pay in certain circumstances, but only a lower amount than other people, and may even be unwilling to pay at all.[66] On the other hand, no account is taken of the willingness to pay of those people who benefit only indirectly, rather than directly, from the reduction of risk and would therefore be willing to pay. Among close family members, there is often a willingness to make a certain financial sacrifice to reduce not only one's own risks but also those of another family member.[67] These kinds of willingness to pay, both to the benefit and to the detriment of third parties, are barely considered in the empirical studies – primarily labour and commodity market research – on the VSL.

2.2.3.4 Problem of the Wealth Effect

In a country-comparison of empirical analyses of the VSL, Viscusi and Aldy attempted to show that the higher a country's per capita income, the higher the value of a statistical life.[68] They note that there is a positive income elasticity between income level and VSL of the order of magnitude of 0.5–0.6.[69] In countries with higher levels of prosperity, the VSL is thus also comparatively higher. Evidence for this is supplied by the empirical findings on the VSL in emerging economies such as India or Taiwan, where the value of a statistical life is significantly lower than in industrial countries like, for instance, the USA.[70] Due to the positive income elasticity, the VSL varies from country to country depending on the wealth status of each particular society.[71]

This wealth effect is easily explained when one takes cognizance of how the willingness-to-pay method is used to determine the VSL. The definitive considerations are the balancing of money and risk on the individual level.[72] Since ability to pay is a precondition of willingness to pay, the choice between money and risk depends very much on the income level of the individual concerned.[73] Thus richer people, having a better ability to pay, are more willing to spend more money on

---

[65] Ashenfelter, p. C17.

[66] Merten, p. 183.

[67] Ackerman and Heinzerling, p. 69; Sunstein, 'Plea', p. 433.

[68] Viscusi and Aldy, pp. 36 ff.

[69] Viscusi and Aldy, p. 40. Accordingly security is a normal good, for which there is rising demand as income rises. See Viscusi, 'Wealth', p. 415.

[70] Merten, p. 206; Sunstein, 'Plea', pp. 414 f.

[71] Sunstein, 'Plea', p. 414 and p. 444.

[72] See Section 2.2.1 above.

[73] Landefeld and Seskin, p. 555; Sunstein, 'Plea', p. 414.

additional safety equipment in their cars than poorer people. Furthermore, individuals with greater human capital and assets have a tendency to pursue low-risk occupations, i.e. their demand for security in the workplace grows.[74] Consequently, wealthy people seem to have a greater risk-aversion, which leads them to place a higher value on the reduction of mortality risks. Accordingly, they are more willing to pay more in order to reduce such a risk of mortality.[75] Ultimately, this greater willingness to pay on the part of rich people also results in a higher VSL. This wealth effect must be given due consideration in the calculation of the VSL by referring to data material that is as representative as possible of the whole of society.

2.2.3.5 Problem of Heterogeneity

Numerous studies on the VSL have been undertaken to date in different countries worldwide, most of which originate from the USA. The results of these various studies were summarized by Viscusi and Aldy in the year 2003 in their comprehensive meta-analysis. The summary makes it clear that among the results of the individual studies, considerable variance is found in the level of the VSL. The average risk factor on which the various studies were based has a substantial influence on these varied results.[76] For example, studies in which employees with relatively risky occupations are overrepresented result in a comparatively low VSL, because these people – as we have seen[77] – often have lower risk-aversion and therefore also demand comparatively lower salary compensation for the acceptance of risk. In contrast, the VSL obtained from studies based primarily on data from less dangerous occupations tends to be at the higher end of the collective results.[78] Furthermore, Viscusi and Aldy found that commodity market studies tended to lead to a somewhat lower VSL than labour market studies.[79]

These considerable divergences between the results of individual studies are fuelled by the lack of a uniform calculation method.[80] Therefore one might question how much this concept is actually worth in practice. For if the value of a statistical life cannot be ascertained reasonably indisputably, it is probably of limited applicability for addressing questions of policy. In response to this objection, however, the varying levels of these results make it clear once again that the VSL ascertained by means of the observable preference statements of the persons concerned are not

---

[74]Garen, p. 9.

[75]Viscusi, 'How', p. 314.

[76]Merten, pp. 192 f.

[77]See Section 2.2.3.2 above.

[78]Viscusi and Aldy, p. 18.

[79]Viscusi and Aldy, p. 24.

[80]Gerner-Beuerle, p. 13. These inadequacies are at least partially addressed in that the majority of studies do not focus on a specific group of people but carry out a comprehensive study across the whole of society. In order to take appropriate account of statistical outliers – both upward and downward – the VSL is determined with reference to the median as well as the mean value from the surveys.

universally valid but only ever relate to the particular risk-laden job under analysis – that is, to a precisely specified risk.[81] It does not therefore give *the* value of a human life as a constant figure.[82] Nor is it necessary, moreover, to talk about *the* value of a human life. Rather, the different values used as a basis by the different regulatory authorities demonstrate that these values, too, may vary or be applied flexibly in different domains. Depending on the risk being addressed, then, there are different VSLs which always need to be understood in their given social context.[83]

2.3 Interim Conclusion

The problematic aspect of the human-capital method is that it bases the valuation of human life solely on the income or tax dues of the individual concerned. It therefore offers no solution for the valuation of humans who have not yet started work or are no longer working (such as children, pensioners, the unemployed, etc.). This method also fails to take account of work done on an unpaid basis. Likewise, the willingness-to-pay method has a number of conceptual and practical shortcomings. In particular, a striking amount of variance is found in the level of the VSL across the different studies. This is primarily because this method is based on a subjective decision by the individuals concerned, and these decisions may differ enormously, not only from one person to another but also, over time, for one and the same person.[84]

The defining difference between the two valuation approaches relates to their contrary evaluative standpoints: while the human-capital method asks quasi *ex post* how much the *death of a human life* costs, the willingness-to-pay method inquires *ex ante* into the value of *avoidance of a mortality.* It is ultimately this difference in conceptual perspective which results in the difference in practical application, i.e. the fact that in the domain of state regulatory interventions for risk reduction, the willingness-to-pay method has superseded the human-capital method. Nowadays the latter remains in use primarily in the courts to assess the damages for loss of a breadwinner in liability cases.

---

[81] Viscusi, 'Survey', p. 203.

[82] Merten, p. 192.

[83] Ackerman and Heinzerling, p. 71; Sunstein, 'Plea', p. 416.

[84] As we have seen, willingness to pay for the reduction of risks and thus the level of the VSL increases in step with income, while rising income is part and parcel of advancing age. Conversely, however, willingness to pay declines in the event of sudden unemployment. Similar changes in willingness to pay over the life-cycle are engendered by an increase in risk-aversion, prompted for example by an increase in responsibility which can occur after marriage or the birth of a child. See Viscusi, 'Devaluation', pp. 106 f.

3 The Moral and Legal Point of View

Leaving aside the aforementioned theoretical and practical shortcomings of the different methods for determining the value of a human life, the concept itself has some fundamental ethical and legal arguments to contend with. For even presuming that there is no alternative to a valuation of life for the purposes cited, the question remains; is it conscionable to place a monetary value on human life at all?

As we have seen, in everyday life, humans constantly take certain risks or accept them as part of life. In doing so, they implicitly express that they do not consider their lives to have infinite value. The critical question is whether the state is entitled to take advantage of such decisions made on an individual level? When it comes to answering this question, essentially there are two opposing scientific camps. On one side are the economists, for whom a cost-benefit analysis – even in respect of human life – is indispensable in the domain of state regulatory interventions and therefore justified. This view is challenged on the other side by a broadly-held ethical conviction that considers the value of human life to be infinite and therefore declares any monetarization to be unacceptable.

In the following, the first question to be pursued is whether human life may be quantitatively valued at all, or whether its monetarization is unacceptable. If the first is answered in the affirmative, this prompts the supplementary question of whether – even with reference to qualitative characteristics – some human lives may be valued more highly than others.

3.1 Permissibility of the Quantitative Valuation of Human Life

To assess the permissibility of the quantitative evaluation of human life, the question to be pursued is whether human life represents a fungible good which may be allocated a monetary value, or whether instead the value of human life is not in fact infinite and therefore defies any form of monetarization. In answering this question, the prohibition against quantification, still to be discussed below, is of central importance.

3.1.1 The Ideal of the Infinite Value of Human Life

Various ethical and religious worldviews – particularly the Western Christian tradition[85] – hold a belief in the "sanctity of life".[86] This belief makes the claim that every human life – regardless of any physical or mental deficiencies – is beyond price, and cannot be weighed in the balance against other human lives and certainly

[85]From the viewpoint of Christian ethics, the human being is made in the likeness of God *(imago dei)*. Therefore human life is sacred and inviolable, and every person possess an absolute and inalienable intrinsic value. See Pope John Paul II, pp. 66 ff.

[86]Ackerman and Heinzerling, p. 67.

not against material commodities. Taken to its ultimate conclusion, it is inferred from this inviolability of human life that every human being is of absolute and infinite worth.[87]

This religious idea of the "sanctity of life" and the consequent claim that human life has infinite value finds its secular justification in the Age of Enlightenment in the worldview of Immanuel Kant, author of the famous declaration:

> In the kingdom of ends everything has either Value or Dignity. Whatever has a value can be replaced by something else which is *equivalent*; whatever, on the other hand, is above all value, and therefore admits of no equivalent, has a dignity.[88]

In the field of human endeavour, then, Kant differentiates between that which has a value (a price) and that which has a dignity. Having a price means being replaceable by something else, whereas having a dignity means being an end in itself. Kant sees humans as rational beings, with moral autonomy that renders them capable of determining their own ends. Therefore they properly have a dignity that "admits of no equivalent".

Drawing on Kant, human dignity and the inviolability of human life have been anchored in most democratic legal systems, and these two legal institutions are instrumental to the justification of the so-called "prohibition against quantification"[89] in legal theory. That is to say, a human life does not admit of quantitative measurement due to its inherent core of human dignity, because the life of the person concerned would thereby be degraded to a mere numerical operand or commodity. Instead, the value of every human life and every moment of life – even for the dying – is infinitely high.[90] In the meantime, this prohibition against the quantification of human life has been embraced by jurisprudence and translated into valid law. Thus the German Federal Constitutional Court in particular has held that the life of one human being may not be weighed against the life of another. The Court ruled that this prohibition flows directly from the core content of the right to life and the guarantee of human dignity, which is why any balancing of a human life represents a violation of human dignity.[91] Thus, the ideal of the infinite value of human life has eventually found enactment in law.

[87] Brech, pp. 209 ff.

[88] "Im Reiche der Zwecke hat alles entweder einen Preis, oder eine Würde. Was einen Preis hat, an dessen Stelle kann auch etwas anderes, als Äquivalent, gesetzt werden; was dagegen über allen Preis erhaben ist, mithin kein Äquivalent verstattet, das hat eine Würde." Kant, p. 434; quoted in English after the translation by Thomas Kingsmill Abbott.

[89] The prohibition on quantification is originally a concept from penal law and states that abstract legal balancing of the protected interest of life is impermissible. Accordingly it is impermissible to weigh one life against another, either numerically or qualitatively, also referred to as the prohibition against the (utilitarian) balancing of human life.

[90] Baumann, pp. 858 f.

[91] BVerfGE 39, 1, pp. 58 f.; last confirmed in BVerfGE 115, 118, pp. 153 f.

3.1.2 The Reality of Value and Resource Scarcity – Or the Normative Force of Actuality

If the view discussed above of the infinite value of human life is, in fact, understood as an ideal to be pursued, then both the state and private actors should do all that is conceivably possible and dedicate the totality of available resources to reducing the risks of mortality and saving human lives. But this claim is obviously disproved in reality – both by the reality of life and the legal reality. With regard to the reality of life, one only need recall the earlier discussion of the acceptance of risk in modern society[92] and the fact that individuals take part in dangerous leisure pursuits – such as mountaineering, paragliding, base-jumping, river-rafting, canyoning, etc. – in which fatal accidents repeatedly occur.[93] This actually observable risk-acceptance makes it clear that individuals do not place an infinitely high value on their own lives.

Likewise, the legal reality refutes this ideal of the infinite value of human life. If life were actually of infinite value, then not only a deliberate fatal shot from a police marksman but also any form of individual assisted dying, or of killing in self-defence or emergency, would be unconstitutional. Equally, no fire-fighter, police officer or soldier could be expected to put his life in danger or be exposed to any risk of death in the course of his duties, since he cannot consent to his own killing by any effective legal means.[94]

The idea of the "sanctity of life" and of the consequent infinite value of human life are, at best, abstract claims (normative postulates),[95] but do not conform to reality and are therefore worthless for practical purposes.[96] From this it is clear that while human lives cannot have infinite value, the question remains as to whether they can be assigned a monetary value or whether in fact they are simply unquantifiable.[97] In answering this question, the distinction between monetary and normative value, which is about to be discussed, is of crucial importance.

3.1.3 Distinction Between Monetary and Normative Value

As the above considerations have shown, the religious and moral idea of the infinite value of human life finds little confirmation in the life of society – on the contrary: day after day, human lives are traded off against one another in politics, society and business. Thus, for example, studies are commissioned by policymakers

---

[92]See Section 2.2.1 above.

[93]Nida-Rümelin, pp. 894 f.

[94]For further examples on the relativization of the infinite value of human life, or the absolute protection of life in law, see Brech, pp. 217 ff.

[95]Even in the Bible there are passages that call into question the "sanctity of life" and hence the prohibition against quantification. Thus for instance, St. John's Gospel, Chapter 11, Verse 50: "[I]t is expedient for us that one man should die for the people, and that the whole nation perish not."

[96]Brech, p. 225; Petersen, p. 437.

[97]Thus for example Ackerman and Heinzerling, p. 67.

in the health-care system to determine whether and to what extent certain medical interventions, which promise to save human lives, are actually cost-effective or financially acceptable to the public; i.e. the cost of the intervention in question is measured against the expected utility of the human lives it hopes to save. Amid these cost-benefit considerations, the human life concerned is at least implicitly attributed a certain monetary value. It is, however, decisive that this monetary "value" is not to be understood as a normative or moral value, but rather a cost point or loss item for the reduction of very tiny risks – in other words, it represents a price. Accordingly, this price of a good cannot be equated with its normative value. For the normative value of an item is something immaterial, and to that extent intrinsically defined, and may also incorporate a certain affection value (*pretium affectionis*). The price, in contrast, is the value of a good expressed in units of money, something material, which is determined by the market and hence extrinsically. Consequently, the VSL represents the actual market value (*vera rei aestimatio*) for the reduction of certain fatal risks, which can be determined by means of the willingness-to-pay method.

3.2 Permissibility of the Qualitative Valuation of Human Life

Having stated earlier that the monetary valuation of – at least statistical – human lives appears to be permissible in the quantitative respect, the focus turns to the question of whether certain differentiations may be made in the level of the VSL based on particular characteristics – such as age, income, etc. Since the impacts of particular regulatory interventions on different population groups can vary, individual public authorities have attempted to make corresponding differentiations and adjustments to the VSL which take account of such characteristics.

Critics raise the objection that this procedure runs counter to conventional conceptions of justice. In particular, they allege that it breaches the principle of equality of rights and the prohibition on discrimination, since it makes reference to a proscribed characteristic. Furthermore, they point out a certain analogy with the slave trade, in which different prices were charged for the purchase of slaves depending on their characteristics, qualities and abilities.[98] If any at all, then only one standard of value, applicable to every human being, ought to be defined. The next line of inquiry is therefore whether it is also permissible to value human life in terms of qualitative aspects; subsequently interest will be focused on differentiation by age and income.

3.2.1 Differentiation by Age

When a selection is made from among different regulatory interventions, it must be taken into account that their effects on the population will not always be equitable. Some such projects tend to favour the younger population whereas others work more

[98] Ackerman and Heinzerling, pp. 63 f.

to the advantage of older people. An optimal allocation of resources demands that the available resources be deployed to wherever they generate the greatest utility for society. In consideration of the fact that measures to reduce mortality risks do not make the beneficiaries immortal but only prolong their lives in statistical terms, the question that arises is whether the valuation of human life should also take account of an individual's age or life expectancy.

In relation to the "Clear Skies Act" mentioned in the Introduction,[99] a regulatory intervention with varying impacts on the survival rates of different age groups, the US Environmental Protection Agency (EPA) had adjusted the value of a statistical life with reference to age: by incorporating the assumption of a "senior discount", the VSL of older people (over 65 years) in comparison to younger people was lowered by 37% from US$ 6.1 million to US$ 3.8 million, a difference of US$ 2.3 million.[100] This quantitative evaluation of life had provoked strong public reactions, since the cut-off point of 65 years of age effectively divided society into two classes. In the face of growing resistance – including from the political sphere – the EPA eventually distanced itself from any differentiation of that kind.[101]

But it is not just that such a differentiation is dubious on ethical and legal grounds; from an economic perspective, the "senior discount" does not correspond to people's willingness to pay for the reduction of small risks. Various studies have shown that, considered over the entire life-cycle, willingness to pay initially rises strongly, reaches its peak in middle age, and finally declines slightly in old age. On this analysis, the VSL follow an inverted U-shaped curve, the initial incline of which is steeper than the subsequent decline after passing its peak.[102]

In view of these insights, the relevant literature proposes instead the *valuation of individual life-years* as an appropriate means of taking life expectancy into account.[103] Having first calculated the *"value of a statistical life-year"*, a fixed average age of death is assumed and the remaining years of life are added. The final total is taken to reflect the value of the human life in question. If this proposal is followed, when assessing the cost-efficiency of a certain regulation, the definitive criterion is not the number of lives to be saved but the number of life-years. As a consequence, preference is given to projects that tend to benefit the younger population and generate lasting utility.[104]

Even if the valuation of single life-years sounds plausible at first, here once again the problem arises that the assumption of a fixed and standard value of a life-year does not conform to the population's real willingness to pay over the course of their

---

[99]See Section 1 above.

[100]Viscusi, 'Devaluation', p. 110.

[101]Sunstein, 'Lives', p. 207.

[102]Sunstein, 'Plea', p. 412; Viscusi, 'Value', p. 589.

[103]Thus Sunstein, 'Lives', p. 207. If health status (quality of life) is considered as well as age, then the measure is known as a "*quality-adjusted life year*". See for example Richard A. Hirth et al.: 'Willingness to Pay for a Quality-adjusted Life Year: In Search of a Standard'.

[104]Sunstein, 'Lives', pp. 249 f.

lifespan, since people of advancing years – partly because of their greater accumulation of assets – tend to spend more money on safety measures and engage in less dangerous activities than when they were younger.[105] In order to reflect this fact, one would have to assign a particular value to each individual life-year, taking an average willingness-to-pay figure for individuals of each age, e.g. US$ 80,000 for the twentieth year of life or US$ 100,000 for the fiftieth. However, this poses a problem to which none of the known methods of economic analysis has yet provided a satisfactory solution.[106]

Thus, until some method has been developed for taking appropriate consideration of age in the valuation of human life, even from an economic perspective a standard, non-age-dependent value should remain the working assumption. For state projects affecting a particular age-group, however, age-related adjustments of the VSL would be possible.[107]

3.2.2 Differentiation by Income

Particularly in relation to the human-capital method, the accusation of discrimination cannot be dismissed out of hand, since in this case – as we have already noted[108] – the life of one concrete person may be valued more or less highly depending on income or tax dues. Nowadays, however, this method barely plays a role in the context of state regulatory interventions, but is only used for calculating damages for the loss of a breadwinner in liability law. In that context, owing to the principle of corrective justice (compensation for damage), differentiations relating to employment are very much in order. Therefore objective grounds are given which justify differential valuations of human life or of damages for the loss of a breadwinner.

Under the willingness-to-pay method, income has an indirect influence on the level of the VSL. As we have seen above,[109] willingness to pay depends greatly on ability to pay, i.e. there is a positive income elasticity between income level and VSL. The conventional calculation of the VSL is based on the assumption of an average willingness to pay, which results in a standard value that is applied equally to all the individuals concerned. This has prompted some to insist that differences between certain individuals' willingness to pay must be reflected in the level of the VSL. Instead of assuming a standard of value, therefore, it is thought that the VSL should be individualized by income, i.e. varied to take account of the individuals' differing income levels.

In practice this means that where individual regulatory interventions favour one particular population group, reference must be made to whichever VSL represents

---

[105] Viscusi, 'Devaluation', p. 112.

[106] Gerner-Beuerle, pp. 13 f.; Viscusi, 'Value', p. 589.

[107] Viscusi, 'Devaluation', p. 113.

[108] See Section 2.1.2 above.

[109] See Section 2.2.3.4 above.

this specific group's concrete average willingness to pay. This may apply predominantly to projects of a local and geographically limited nature which favour one precisely specified group, such as state-financed security measures in an up-market district.[110] Also conceivable are state-supported services which tend to be used more by people on higher incomes, such as airport security controls.[111]

The accusation of discrimination is often levelled at such income based differentiations, since they imply that the rich might be of greater value than the poor. One reply to this accusation may be that this is not a matter of the value of life, but of willingness to pay for the reduction of small risks of mortality.[112] Furthermore, say its opponents, people with higher incomes pay more taxes towards the funds from which state projects are ultimately financed. The net contribution of a wealthy person to the financing of these projects is therefore inequitably higher than that of a poorer person. In the absence of income-related differentiations in this area, they argue, the final outcome would be a redistribution of income.[113] Ultimately they believe that those who benefit most from a collective project should bear its cost.[114] The counterargument, however, is that universal taxation is based on the principle of capacity to pay, not the principle of equivalence.

3.3 Interim Conclusion

In summary, it can be stated that the idea of the infinite value of human life does not match up to the reality of life, nor to legal reality. Instead, humans repeatedly enter into fatal risks, thereby implicitly expressing that they cannot place an infinitely high value on their own lives. Richard A. Posner aptly makes this point:

> [I]f [. . .] the value of life is always infinite, then [. . .] people would never take any risks – an obviously false description of human behaviour.[115]

In addition, not only does the general scarcity of resources make it impossible to guarantee the absolute protection of life, but a risk-free society appears to be neither desirable nor financially affordable. In other words, from a normative perspective human lives may have infinitely high value, but they nevertheless have a price when their safety is at stake. However, this price does not express the value of a real person but that of a statistical person. The level of this value may vary depending on the domain of application and the group of addressees, since it is based on the willingness to pay of the people whom the measure is intended to benefit. Within

110 Sunstein, 'Plea', p. 415.

111 Viscusi, 'Survey', p. 213.

112 Sunstein, 'Plea', p. 444.

113 Viscusi, 'Survey', pp. 212 f.

114 Sunstein, 'Plea', p. 445.

115 Posner, p. 198.

this group of addressees, however, it is average willingness to pay that determines the value of a statistical person.

4 Conclusion

When *concrete* human lives are at stake – e.g. miners trapped in a collapsed mine[116] – there is a universal taboo against any refusal to mount a rescue action on the grounds that the costs of rescue efforts would exceed the value of the human lives concerned. In the case of *abstract* risks in everyday life, on the other hand, we tacitly accept a certain risk of mortality and thereby implicitly attribute a value to life. For example, we continue to lay out boulevards along busy roads although the resulting restriction of drivers' vision considerably increases the probability of fatal traffic accidents. Thus, greater weight is implicitly attached to the utility of aesthetic landscaping than to the value of the sacrificed lives. Does not this differentiation between concrete and abstract endangerment of life pose a valuation conflict? Evidently we find it easier to sacrifice mere *statistical* lives than *concrete* lives, although we know that this *ex ante* abstract mortality risk will, at some point, become a reality and eventually claim a concrete accident victim.[117] This brings to light an obvious anomaly of human behaviour, since we humans are prepared to devote far greater costs to concrete rescue efforts than to preventative safety measures for risk reduction, despite the fact that in the long run the latter could save far more lives at lower cost overall.[118]

Hence, when assessing the cost efficiency of such state regulatory interventions, the corresponding cost-benefit considerations always refer to future, yet to occur, but actually possible deaths. Therefore what is calculated is not the valuation of a concrete human life – as is often misunderstood by critics of this concept – but of a statistical life of those potentially affected by the intervention in question. For this reason, the usual concept quoted in the relevant literature is the "value of a *statistical* life".[119]

This terminological distinction ultimately makes it possible to defuse the debate between economists and philosophers. The latter strictly reject any valuation of human life, due to their ethical conviction that the value of a human life is infinite, and therefore that any monetarization of human life is indefensible on moral grounds. This view is based on a misapprehension, however, since the economic valuation of human life is not about the monetary value of a concrete human life

---

[116] As occurred in the summer of 2010 in the Chile mining disaster, when the collapse of a mine near San José trapped 33 miners. As if by a miracle, all survived unharmed and were finally rescued successfully after months of preparation and the expenditure of vast sums.

[117] Huster and Kliemt, pp. 248 and 250.

[118] Fried, p. 1416. The main reasons for this preference seem to be the identifiability of and personal bond with the victim, i.e. the person in concrete danger. On this, see Fried, pp. 1428 ff.

[119] Viscusi, 'How', p. 322.

but the monetary valuation of the reduction of mortality risks. Richard A. Posner's account of this fine but crucial distinction is apposite:

> So while valuing human lives sounds like an ethical or even a metaphysical undertaking, what actually is involved is determining the value that people place on avoiding small risks of death and the transformation into a 'value of life' is done simply to determine how large the award of damages in a lump sum must be in order to be equivalent.[120]

Thus, the debate between philosophers and economists over the permissibility of placing a value on human life is purely a terminological problem, which can easily be resolved in this context by talking about valuation of the reduction of mortality risks instead of the value of life.[121]

At the same time, this terminological adjustment forestalls the reproach of a *naturalistic fallacy* which is often levelled at those economists who derive a *normative* claim from the purely *descriptive* results of empirical data surveys. For the economic studies on the VSL give a purely *descriptive* account of the individual willingness to pay for the prevention of one death. From these factual realities (the "is") it is not possible to derive any *normative* claim (an "ought") about a finite value of human life as a moral principle.[122] Philosophers and jurists for their part must, however, be careful not to inadvertently commit a *normativistic fallacy* by drawing conclusions from the postulate of the infinite value of human life (an ought-proposition) about the reality of life (an is-proposition).[123]

Therefore it is vital for the relevant regulatory authorities that the VSL is understood to be purely descriptive and not normative. It should serve only as an instrument for assessing the cost-efficiency of risk-laden state projects, but should not be the only criterion by which the realization of the project in question is decided. The relevant regulatory authorities are therefore entreated to understand the concept of the VSL for what it is; namely a figure in the cost-benefit analysis, which makes no normative claims of any kind and to which no normative significance may be imputed.

Acknowledgment I would like to thank my mentor, Prof. Dr. Klaus Mathis, for his insightful advice and continuing support.

Bibliography

Ackerman, Frank and Heinzerling, Lisa, *Priceless: On Knowing the Price of Everything and the Value of Nothing* (New York, 2004)

Ashenfelter, Orley, 'Measuring the Value of a Statistical Life: Problems and Prospects', in *The Economic Journal*, Vol. 116/510 (2006), pp. C10 ff.

---

120 Posner, p. 197.

121 Posner and Sunstein, p. 557.

122 Mathis, 'Efficiency', p. 31.

123 On the normativistic fallacy see Petersen, pp. 436 f.

Baranzini, Andrea and Ferro Luzzi, Giovanni, 'The Economic Value of Risks to Life: Evidence from the Swiss Labour Market', in *Schweizerische Zeitschrift für Volkswirtschaft und Statistik*, Vol. 137 (2001), pp. 149 ff.

Baumann, Karsten, 'Das Grundrecht auf Leben unter Quantifizierungsvorbehalt? Zur Terrorismusbekämpfung durch "finalen Rettungsschuß"', in *Die Öffentliche Verwaltung*, Vol. 57 (2004), pp. 853 ff.

Brannon, Ike, 'What Is a Life Worth? Despite Its Prima Facie Callousness, Determining the Value of a Human Life Is Necessary for Good Public Policy', in *Regulation*, Vol. 27/4 (2004), pp. 60 ff.

Brech, Alexander, *Triage und Recht. Patientenauswahl beim Massenanfall Hilfebedürftiger in der Katastrophenmedizin. Ein Beitrag zur Gerechtigkeitsdebatte im Gesundheitswesen* (Berlin, 2008)

Broome, John, *Weighing Lives* (Oxford and New York, 2004)

Calabresi, Guido, 'The Decision for Accidents: An Approach to Nonfault Allocation of Costs', in *Harvard Law Review*, Vol. 78 (1965), pp. 713 ff.

Calabresi, Guido, *The Costs of Accidents: A Legal and Economic Analysis* (New Haven, 1970)

Eeckhoudt, Louis R. and Hammitt, James K., 'Does Risk Aversion Increase the Value of Mortality Risk?', in *Journal of Environmental Economics and Management*, Vol. 47 (2004), pp. 13 ff.

Fried, Charles, 'The Value of Life', in *Harvard Law Review*, Vol. 82 (1969), pp. 1415 ff.

Garen, John, 'Compensating Wage Differentials and the Endogeneity of Job Riskiness', in *The Review of Economics and Statistics*, Vol. 70 (1988), pp. 9 ff.

Gerner-Beuerle, Carsten, 'Recht, Effizienz und Calabresis Trugschluss', in Stefan Grundmann et al. (eds.), *Unternehmensrecht zu Beginn des 21. Jahrhunderts. Festschrift für Eberhard Schwark zum 70. Geburtstag* (Munich, 2009), pp. 3 ff.

Griffin, James, 'Are There Incommensurable Values?', in *Philosophy and Public Affairs*, Vol. 7 (1977), pp. 39 ff.

Hirth, Richard A. et al., 'Willingness to Pay for a Quality-adjusted Life Year: In Search of a Standard', in *Medical Decision Making*, Vol. 20 (2000), pp. 332 ff.

Huster, Stefan and Kliemt, Hartmut, 'Opportunitätskosten und Jurisprudenz', in *Archiv für Rechts- und Sozialphilosophie*, Vol. 95 (2009), pp. 241 ff.

Kant, Immanuel, *Grundlegung zur Metaphysik der Sitten*, in Kants Werke, Akademie-Textausgabe, Vol. IV (Berlin, 1968), pp. 385 ff., quoted in English after the translation by Thomas Kingsmill Abbott, *Fundamental Principles of the Metaphysics of Morals* (New York, 2009)

Landefeld, Steven J. and Seskin, Eugene P., 'The Economic Value of Life: Linking Theory to Practice', in *American Journal of Public Health*, Vol. 72 (1982), pp. 555 ff.

Mathis, Klaus, *Efficiency instead of Justice? Searching for the Philosophical Foundations of the Economic Analysis of Law* (New York, 2009; cited as: *Efficiency*)

Mathis, Klaus, 'Zukünftige Generationen in der Theorie der Gerechtigkeit von John Rawls', in Sandra Hotz and Klaus Mathis (eds.), *Recht, Moral und Faktizität. Festschrift für Walter Ott* (Zurich and St. Gallen, 2008), pp. 181 ff. (cited as: 'Generationen')

Merten, Carsten, *Die Bewertung menschlichen Lebens im Haftungsrecht* (Frankfurt a.M., 2007)

Mishan, Ezra J., 'Evaluation of Life and Limb: A Theoretical Approach', in *Journal of Political Economy*, Vol. 79 (1971), pp. 687 ff.

Mishan, Ezra J. and Quah, Euston, *Cost-Benefit-Analysis* (5th edn., London, 2007)

Nida-Rümelin, Julian, 'Wert des Lebens', in Julian Nida-Rümelin (ed.), *Angewandte Ethik. Die Bereichsethiken und ihre theoretische Fundierung. Ein Handbuch* (2nd edn., Stuttgart, 2005)

Petersen, Niels, 'Braucht die Rechtswissenschaft eine empirische Wende?', in *Der Staat*, Vol. 49 (2010), pp. 435 ff.

Pope John Paul II, *"Evangelium vitae". Encyclical Letter on the Value and Inviolability of Human Life. March 25, 1995* (Rome, 1995)

Posner, Richard A., *Economic Analysis of Law* (7th edn., New York, 2007)

Posner, Eric A. and Sunstein, Cass R., 'Dollars and Death', in *The University of Chicago Law Review*, Vol. 72 (2005), pp. 537 ff.

Schelling, Thomas C., 'The Life You Save May Be Your Own', in Samuel B. Chase (ed.), *Problems in Public Expenditure Analysis* (Washington DC, 1966)
Simon, Herbert A., 'A Behavioral Model of Rational Choice', in *Quarterly Journal of Economics*, Vol. 69 (1955), pp. 99 ff.
Smith, Adam, *An Inquiry into the Nature and Causes of the Wealth of Nations*, edited by Edwin Cannan (New York, 1937)
Spengler, Hannes, 'Kompensatorische Lohndifferentiale und der Wert eines statistischen Lebens in Deutschland', in *Zeitschrift für ArbeitsmarktForschung*, Vol. 37 (2004), pp. 269 ff.
Sunstein, Cass R., 'Bad Deaths', in *Journal of Risk and Uncertainty*, Vol. 14 (1997), pp. 259 ff. (cited as: 'Deaths')
Sunstein, Cass R., 'Lives, Life-Years, and Willingness to Pay', in *Columbia Law Review*, Vol. 104 (2004), pp. 205 ff. (cited as: 'Lives')
Sunstein, Cass R., 'Valuing Life: A Plea for Disaggregation', in *Duke Law Journal*, Vol. 54 (2004–2005), pp. 385 ff. (cited as: 'Plea')
Viscusi, Kip W., 'Wealth Effects and Earnings Premiums for Job Hazards', in *Review of Economics and Statistics*, Vol. 60 (1978), pp. 408 ff. (cited as: 'Wealth')
Viscusi, Kip W., 'The Value of Life in Legal Contexts: Survey and Critics', in *American Law and Economics Review*, Vol. 2 (2000), pp. 195 ff. (cited as: 'Survey')
Viscusi, Kip W., 'The Value of Life', in Steven N. Durlauf and Lawrence E. Blume (eds.), *The New Palgrave Dictionary of Economics*, Vol. 8 (2nd edn., New York, 2008), pp. 586 ff. (cited as: 'Value')
Viscusi, Kip W., 'How to value a life?', in *Journal of Economics and Finance*, Vol. 32 (2008), pp. 311 ff. (cited as: 'How')
Viscusi, Kip W., 'The Devaluation of Life', in *Regulation and Governance*, Vol. 3 (2009), pp. 103 ff. (cited as: 'Devaluation')
Viscusi, Kip W. and Aldy, Joseph E., 'The Value of a Statistical Life: A Critical Review of Market Estimates throughout the World', in *Journal of Risk and Uncertainty*, Vol. 27 (2003), pp. 5 ff.
Viscusi, Kip W., Harrington, Joseph E. (Jr.) and Vernon, John M., *Economics of Regulation and Antitrust* (4th edn., Cambridge MA, 2005)

Index

K. Mathis (ed.), *Efficiency, Sustainability, and Justice to Future Generations*,
Law and Philosophy Library 98, DOI 10.1007/978-94-007-1869-2,

MIX
Papier aus verantwortungsvollen Quellen
Paper from responsible sources
FSC® C105338

If you have any concerns about our products,
you can contact us on
ProductSafety@springernature.com

In case Publisher is established outside the EU,
the EU authorized representative is:
Springer Nature Customer Service Center GmbH
Europaplatz 3, 69115 Heidelberg, Germany

Printed by Libri Plureos GmbH
in Hamburg, Germany